T0338296

Seafood Chilling, Refrigeration and Freezing

Seafood Chilling, Refrigeration and Freezing

Science and Technology

Nalan Gökoğlu and Pınar Yerlikaya

Fisheries Faculty, Akdeniz University, Antalya, Turkey

WILEY Blackwell

This edition first published 2015 © 2015 by John Wiley & Sons, Ltd.

Registered Office
John Wiley & Sons, Ltd., The Atrium, Southern Gate, Chichester, West Sussex, PO19 8SQ, UK

Editorial Offices
9600 Garsington Road, Oxford, OX4 2DQ, UK
The Atrium, Southern Gate, Chichester, West Sussex, PO19 8SQ, UK
111 River Street, Hoboken, NJ 07030-5774, USA

For details of our global editorial offices, for customer services and for information about how to apply for permission to reuse the copyright material in this book please see our website at www.wiley.com/wiley-blackwell.

Library of Congress Cataloging-in-Publication Data

Gökoğlu, Nalan, author.
  Seafood chilling, refrigeration and freezing : science and technology / Nalan Gökoğlu and Pınar Yerlikaya.
    pages cm
  Includes bibliographical references and index.
  ISBN 978-1-118-51218-0 (cloth)
1. Frozen seafood. 2. Frozen fish. 3. Refrigeration and refrigerating machinery.
I. Yerlikaya, Pınar, author. II. Title.
  SH336.F7G65 2015
  664′.9453–dc23
                                                    2015007742
A catalogue record for this book is available from the British Library.

Wiley also publishes its books in a variety of electronic formats. Some content that appears in print may not be available in electronic books.

Cover image: ice background ©sbayram/istockphoto; three salmon pieces on a chopping board ©olgna/istockphoto; Raw sea bass fish on cutting board top view ©ALLEKO/istockphoto; Fish on ice ©PapaBear/istockphoto.

Set in 10/13.5pt Meridien by SPi Global, Pondicherry, India
Printed and bound in Malaysia by Vivar Printing Sdn Bhd

1   2015

Contents

Preface

Fish and other seafood are the major sources of nutritious protein and micronutrients. They form part of a healthy diet due to their content of high-quality protein with essential amino acids, minerals and vitamins. However, their flesh is perishable feature and causes spoilage. Therefore, preservation of seafood is an important issue. The preservation methods lowering the temperature protect the original properties of these products. The first application on board a vessel is chilling or freezing. These preservation methods are used comprehensively for fish and fish products. Books on chilling, refrigeration and freezing are generally available for all foods, but there is a limited number of books specializing on fish.

In this book, besides general knowledge on chilling, refrigeration and freezing, seafood-specific applications are given. I hope that this book will be useful for researchers, students and industrialists.

The authors would like to thank their families for their support and patience.

Drawings: Dr. Yasar Ozvarol.

Nalan Gökoğlu
and Pınar Yerlikaya

CHAPTER 1

Introduction

1.1 Spoilage of seafood

Fish can be easily spoiled after death. The decomposition of fish flesh occurs mainly due to various chemical, microbial and enzymatic actions. Microorganisms are found on the skin, gill surfaces and in the intestines of live fish. In live fish, these microorganisms do not affect on fish quality due to the normal body defences of fish. However, microorganisms attack fish tissues after death. While numerous microorganisms can cause spoilage of fish, the main ones are bacteria. The bacterial flora of fish is affected by several factors, including season and environment. The bacterial microflora of fish is related to the microbial population of the water in which it lived. Psychrophilic and mesophilic microorganisms are responsible for the fish spoilage. Microorganisms enter the body of fish through gills, blood vessels, skin and abdominal wall. Moreover, bacteria may enter through injured tissues. Bacteria cause undesirable flavour and taste changes in the flesh of fish. Besides flavour and taste, bacteria are responsible for the changes in appearance and physical properties of fish. Deteriorative changes in fish are due to decomposition of non-protein nitrogen compounds. Proteins are degraded into peptides, amino acids, ammonia and some other low-molecular weight

Seafood Chilling, Refrigeration and Freezing: Science and Technology, First Edition.
Nalan Gökoğlu and Pınar Yerlikaya.
© 2015 John Wiley & Sons, Ltd. Published 2015 by John Wiley & Sons, Ltd.

nitrogen compounds. The deteriorative changes occurring in fish result in the gradual accumulation of certain compounds in the flesh. Enzymes remain active after the death of the fish and are particularly involved in flavour changes that take place during the first few days of storage. Autolysis is the breakdown of proteins, lipids and carbohydrates by enzymes. The initial quality loss in fish occurs by these autolytic changes. All of the factors affecting the quality of fish, such as bacteria and enzymes, may bring about sensory changes, which are unacceptable for the consumer.

1.2 Preservation of seafood

Since fresh fish spoil easily, they need to be processed and preserved. Preservation provides a long shelf-life for fish and fish products. Preservation can be defined as the storage of excess fish when they are abundantly caught or produced so they can be consumed as if fresh in times when food is scarce or when transported to long distances. Preservation affects food in two ways: (1) it keeps the original freshness and properties of fish; (2) it changes the original properties of the food and creates new product. The main purpose of both of these is to prevent spoilage, especially by microorganisms. Several preservation methods have been developed, some of them providing a longer shelf-life than others. The choice of a preservation method depends on the product, properties of the product, availability of energy, the storage facilities, and the costs of the method. It is sometimes necessary to combine methods.

Fish spoils very quickly in high ambient temperatures, because chemical, physical and microbiological actions accelerate in high temperatures. Therefore, the temperature should be reduced immediately after harvest. In this regard, preservation begins in fishing vessels for fish and fishery products. The first preventative step to keep the quality of fish is taken onboard. Chilling, refrigeration and freezing are generally used onboard as preservation methods; these methods are also common in inshore applications. The fish are transported to land under cold conditions, and stored in cold storage until processing or marketing in the plant. Products remain fresh under refrigeration for a few days; they can be stored much longer when frozen. Low temperatures must be maintained accurately and continuously.

1.2.1 Chilling

Chilling is to reduce fish temperature to 0°C. The main aim of chilling is to prevent physical, chemical and microbiological activities occurring under normal conditions by reducing the temperature. Chilling cannot completely stop spoilage of fish but retards it. Effective chilling depends on some factors, including initial microbial load, chemical composition, temperature, relative humidity, and air velocity. The lower the temperature means the longer the shelf life. Mesophilic and thermophilic microorganisms are retarded at chilling temperature. Different chilling methods are used for fish and fishery products. The most common and effective method is chilling with ice. In this method the fish is completely surrounded by ice because the cooling capacity of ice is very good. Melting ice removes heat from the fish and so cools it. Moreover, chilled or refrigerated sea water (RSW) is used for chilling of fish. This method is common in onboard applications.

1.2.2 Refrigeration

Refrigeration is also a method of lowering the temperature of the product. In this method mechanical cooling is used. Air is cooled by a refrigerator and cold air is passed over the surface of a fish to rapidly cool it. Air takes the moisture from the surface of the product, and therefore surface of the fish becomes dry. For this reason, refrigeration is more suitable for iced fish. After icing of fish in boxes or containers, they are stored under refrigeration and effective cooling is achieved in this way. On the other hand, frozen products should be stored in cold conditions until use. Different refrigeration systems and refrigerants are used for fish and fishery products. Refrigeration equipment can be installed in fishing vessels. Thus, fish quality keep just after catching. RSW is a good chilling method on board, and refrigerated equipment installed in the vessel produces RSW.

1.2.3 Freezing

Preservation of fish and fishery products for longer periods can be achieved by freezing. Freezing is the process of removing heat to lower product temperature to −18°C or below. It has the advantage of minimizing microbial and enzymatic activity. Microbial and enzymatic activities are limited by lowering temperature and water activity. Many spoilage bacteria can be destroyed by freezing. In order to continue this effect of freezing, the frozen state must be protected. Frozen products must be stored in the cold until use and the cold chain definitely should not be broken.

Thawing is a very important process for frozen seafood. If thawing is not performed in proper conditions, the quality of frozen fish is significantly affected, even if frozen in good conditions. Thawing at low temperatures will prevent the loss of quality of the fish. Several thawing methods are used for fish and fishery products. Whichever method is used, rapid thawing is essential.

In this book; chilling, refrigeration and freezing, which are important preservation methods in fishery and fish industry are defined. Uses of these methods are described individually. These methods, especially chilling and refrigeration are very important because they are applicable after catch onboard. Freezing also is applicable in factory vessels. On the other hand, freezing is the most effective method to preserve the original quality of fish for longer periods. If sensitivity of fish to spoilage is remembered, the importance of these preservation methods will be understood. To extend the shelf life of fish and fish products, even a few hours is very important.

CHAPTER 2

Chemical composition of fish

2.1 Proteins

The major constituent of fish flesh is water, which accounts for about 70–80% of the weight of the fillet. The water in fresh fish muscle is tightly bound to the proteins in the structure. There is an inverse relationship between water and lipid content in fish. During different seasons, with an increase in fat content, there is a decrease in water content. The moisture content is also known to generally decrease with age. The water content of lean fish increases during sexual maturation. Red lateral muscle includes slightly less protein and more lipid than the white muscle. The posterior part of the fish fillet contains more protein and fewer lipids than the anterior part. Lipids are energy reserves and are utilized in the maintenance of life. In case of migration or spawning periods, protein is utilized for energy, in addition to lipids, resulting in a reduction of biological condition.

Proteins are essential nutrients for growth and as constituents of the body's cells. Amino acids play a prominent role as the building materials of proteins. The type and rank order of the amino acids determines the conformational structure, chemical and biological properties

Seafood Chilling, Refrigeration and Freezing: Science and Technology, First Edition.
Nalan Gökoğlu and Pınar Yerlikaya.
© 2015 John Wiley & Sons, Ltd. Published 2015 by John Wiley & Sons, Ltd.

of the protein (Saldamli 1998). All amino acids, except for essential amino acids, are synthesized by transaminase enzyme in the liver and transamination reactions, in which vitamin B6 serves as a coenzyme. Essential amino acids cannot be synthesized by humans and other mammals and hence must be supplied in the diet. Fish is known to be a good source of protein, rich in essential amino acids such as lysine, cystine, methionine, threonine and tryptophan (Usydus *et al.* 2009). The decisive factors of the nutritive quality of protein are the content of essential amino acids, the presence of specific essential amino acids similar to that found in the human body, the energy supplied, and the digestibility of the protein. The ease of digestion of fish is due to the low connective tissue content and the shortness of the muscle fibres. The most important attribute of animal-derived proteins satisfies these features by possessing adequate and balanced essential amino acids.

The crude protein content of seafood ranges from 17 to 22%. In crustaceans and molluscs, protein levels can vary from 7 to 23%. Protein and lipid contents of fish increase just before spawning. Protein content also increases in spring when more food becomes available. Fish and shellfish muscle proteins are classified, based on solubility in salt solutions, into three main groups: such as sarcoplasmic, myofibrillar and stromal proteins (Huss 1995).

2.1.1 Sarcoplasmic proteins

Sarcoplasmic proteins, which can be soluble in water and dilute salt solutions, comprise about 15–30% of the total protein in fish muscle. These proteins consist of hundreds of enzymes, pigmented proteins such as myoglobin and haemogobin, and other albumins. In addition, antifreeze proteins and glycoproteins in fish caught in cold water are included in this group. Unlike land animals, fish contain more Ca^{2+}-binding proteins.

The red muscle of fish has a darker appearance, due to high concentration of myoglobin. Red muscle contains more mitochondria and less sarcoplasmic reticulum than white fibres, which are required for prolonged aerobic metabolism of energy reserves. The muscles of pelagic fish contain significant amounts of dark muscle containing myoglobin, which are equipped for prolonged aerobic activity. Demersal fish do not swim actively for long periods as they tend to drift with ocean currents. The content of sarcoplasmic protein is higher in pelagic fish than in demersal fish. The myoglobin content of muscle increases with age, and during the migration season.

Oxymyoglobin and oxyhaemoglobin are responsible for the colour characteristics of fish muscle. During handling and storage haemoglobin dissolves easily, whereas myoglobin is retained in the cell structure. Some molluscs, crustaceans and certain colourless blood Antarctic fish species, for instance, contain no haemoglobin. Shellfish have copper-containing proteins called haemocyanins.

The edible quality of the fish is determined by hydrolases, oxidoreductases and transferase enzymes. Sarcoplasmic enzymes are responsible for the deterioration of the fish muscle. The presence of sarcoplasmic proteins has an adverse affect on the strength, the deformability of myofibrillar protein gels, and the water-holding capacity. The low gel strength of the products of mackerel and sardine can be explained by their sarcoplasmic protein content.

The content and composition of the sarcoplasmic proteins can vary between species. The electrophoretic patterns of sarcoplasmic protein fractions can be utilized as fingerprints to identify fish species.

2.1.2 Myofibrillar proteins

Myofibrillar proteins are structural proteins that compose 65–70% of the fish muscle protein. They are soluble in high salt solutions. The proportion of myofibrillar protein to total muscle protein is higher in fish than in land animals.

Myosin and actin are responsible in muscle contraction–relaxation cycle. In post-mortem muscle, myosin and actin exist as an actomyosin complex. Myosin, ranging from 50 to 60%, forms the thick myofilaments, whereas actin, accounts for 15–20%, is the principal component of the thin filaments. The isoelectric point of myosin is at pH 5.0–5.3 and the actin molecule has an isoelectric point at pH 4.7. The other regulatory proteins are tropomyosin, troponin, actinin, C, I and T proteins. The myosin ATPase activity is required for the interaction of myosin with actin. The formation of actomyosin is blocked by binding adenosine triphosphate (ATP) with myosin in living organisms. Troponin and tropomyosin are also responsible for prevention of actomyosin formation during relaxation. Fish actomyosin has been found to be labile and easily changed during processing and storage. During frozen storage, the actomyosin becomes tougher. Fish myosins are unstable, being more sensitive to denaturation, coagulation, degradation, or to chemical changes (Venugopal 2009).

Myosin and actin are also responsible for important functional properties in food systems, such as water-holding, emulsifying capacity, binding ability and gelation. The rheological and functional properties of fish proteins play a significant role in the preparation of surimi based products. Gel-forming abilities differ among fish species. Cod and silver hake can have the ability of gelatinization comparing to herring due to their cross-linking abilities and forming large protein aggregates by myosin heavy chain (Chan *et al.* 1992).

2.1.3 Stroma proteins

The insoluble matter remaining after removing sarcoplasmic and myofibrillar proteins from muscle is called stroma or connective tissue proteins. They consist predominantly of collagen, with the remainder being elastin and gelatin. Stroma proteins are located in the extracellular matrix, accounting for 3% of the total muscle protein. However, elasmobranch fish such as shark, ray and skate can contain up to 10% stroma proteins. This low content of collagen gives the soft texture to fish meat (Sivik 2000). During chill storage the myocommata of fish may fail to hold the muscle cells together, causing gaping of the flesh. Collagen, in addition to being present in muscle tissue, can also be found as a major structural protein in fish skin, bones and scales. This triple helix protein contains repeated glycine-proline-hydroxyproline-glycine amino acid sequences. The collagen present in fish muscle is rich in essential amino acids and is more thermolabile and contains fewer, but more labile, cross-links than collagen from warm-blooded vertebrates. The thermal alteration of collagen is important in hot smoking process, canning technology, short-time sterilization and in utilization of fish waste. The mantle muscle of some squid species can be tough after cooking because of these thermal changes and the quality changes to fresh and frozen fish after death is the result of collagen alterations.

Proteins are utilized in many industrial applications. They form emulsions with unsaturated fatty acids, in order to generate more stability against oxidation. Fish proteins, including myofibrillar and sarcoplasmic proteins, have been used as film-forming material. Bioactive peptides isolated from various fish protein hydrolysates have shown numerous bioactivities such as antihypertensive, antithrombotic, immunomodulatory and antioxidative activities (Harnedy & FitzGerald 2012). The separation of the muscle constituents is necessary for various physiological and biochemical studies. The gel-forming ability of protein has great importance in products such as surimi and kamaboko, which are consumed willingly in eastern countries, such as Japan, China and Korea. Therefore, the purification and fractionization of myofibrillar proteins have attracted the attention of researchers. Protein concentrates are utilized as food supplements for infants, sportsmen and patients, in order to enrich protein intake, and are applied in various food industries such as gelating or emulsion agents.

2.1.4 Non-protein nitrogen compounds

In addition to proteins, other nitrogenous compounds are present in fish muscle. They are categorized as non-protein nitrogen, including chemical compounds such as amino acids, small peptides, creatine, creatine phosphate, creatinine, amine oxides, guanidine compounds, quaternary ammonium compounds, nucleosides, and nucleotides (including ATP). These compounds are responsible for not only sensorial characteristics but also contribute to the spoilage of fishery products. They are often volatile and malodorous (Sanchez-Alonson *et al.* 2007). The occurrence and properties of proteins and non-protein nitrogen components in fish are the determinants of dehydration, freezing, thermoprocessing and fermentation characteristics (Hargin 2002).

The distribution of these compounds varies with species, freshness and environmental factors. The non-protein nitrogen constituted about 10% of the total nitrogen in teleost fish, 20% in crustaceans and molluscs and over 30% in elasmobranchs (Velankar & Govindan 1958).

2.1.4.1 Free amino acids

The main constituents of flavour compounds in fisheries are amino acids, nucleotides, guanidine compounds and quarternery ammonium compounds. The individual amino acids (such as glycine, valine, alanine, and glutamic acid) are known to contribute to taste, together with the degradation components of nucleotides such as inosine (Olafsdottir & Jonsdottir 2010).

The sweet taste of fresh shrimp and crab is due to their free glycine content. Shrimp, lobster, crab, squid and other shellfish generally contain larger amounts of amino acids, including arginine, glutamic acid, glycine, and alanine, than finfish. The higher contents of these amino acids during the winter season make squids more palatable, as compared with those harvested in summer (Venugopal 2009). Elasmobranchs appear to have higher amount of free amino acid nitrogen content than teleosts (Sen 2005).

Some unique non-protein amino acids such as taurine, β-alanine, methylhistidine and proline dominate in most fish. Taurine contributes to osmoregulation, serves as food reserve and is active in the Maillard browning reaction (Haard 1995). It is also important in neural development. Adult humans can synthesize taurine in a small amount. Molluscs such as mussel and scallops are rich in taurine, meanwhile crabs and some fish species contain less taurine (Spitze *et al.* 2003). The muscles of molluscs and crustaceans are rich in free amino acids. Fish seems to be unique among meat-producing animals in having free histidine in its muscle (Sen 2005). Red muscles tend to contain more histidine than white muscles. The tissues of scombroid fish, such as tuna and mackerel contain high levels of free histidine, which may be converted into histamine by associated microorganisms. The levels of free amino acids usually increase in fishery products during storage due to action of endogenous and exogenous proteases (Gökoğlu *et al.* 2004a).

2.1.4.2 Peptides

Three basic dipeptides are characterized in fish muscle: carnosine (β-alanyl histidine), anserine (β-alanyl-1-methyl histidine) and balenine (β-alanyl-3-methyl histidine), which is a characteristic constituent of whale muscle. Dark muscles tend to contain these compounds more than white muscles. The ratio of carnosine to anserine is higher in freshwater than marine fish. Anserine, as well as carnosine, was reported to have strong ability to eliminate hydroxyl radicals and singlet oxygens (Kikuchi *et al.* 2004).

2.1.4.3 Nucleotides

Most of the nucleotides present in fish muscle are formed by ATP degradation products. In living organisms, muscle contraction is powded by the release of energy during the breakdown of ATP. When the oxygen level is insufficient after death, the muscle tends to shift to anaerobic metabolism. ATP is gradually depleted by membrane and

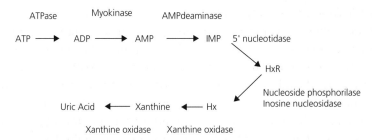

Figure 2.1 Degradation of ATP.

contractile ATPase enzymes and microbial metabolism also contributes to degradation. A series of reactions results in the conversion of ATP through several compounds; ATP is sequentially degraded to adenosine diphosphate (ADP), adenosine monophosphate (AMP), inosine monophosphate (IMP), inosine (HxR), and hypoxanthine (Hx) by autolytic enzymes, as shown in Figure 2.1.

In most fish species, ATP degrades very quickly to IMP, and this compound is reported to be desirable since it has flavour-enhancing properties while the accumulation of Hx is slow and results in an unpleasant taste. The concentrations of ATP and its breakdown products are most widely used as indices of freshness in many fish species. A strong correlation has been observed between nucleotide catabolism and the loss of freshness of fish. Using the ratio of the concentrations of inosine and hypoxanthine to the total amount of ATP-derived compounds – (the K value) – is a good measurement of fish muscle quality (Saito *et al.* 1959).

Degradation of ATP and related nucleotides in frozen fish occurs mainly around $-5°C$ and $-15°C$ and is found less at lower temperatures. Therefore, the measurement of AMP, IMP and Hx is not very suitable for quality determinations of frozen fish (Hedges 2002). Since adenosine nucleotides are almost converted to IMP in the short term, the Ki value, which only excludes ATP, ADP and AMP, is used.

Nicotinamide adenine nucleotide (NAD) is another nucleotide present in fish muscle. NAD and its derivates function as a cofactor in oxidation/reduction; NAD^+ can also be used as a substrate in several biochemical reactions in marine-derived organisms such as Maillard browning and post-harvest pH alterations. Dark muscle contains about twice that in white muscle.

2.1.4.4 Guanidine compounds

The phosphorylated form of creatine plays an important role in fish muscle, acting as an energy reservoir. Creatinine phosphate is rapidly converted to free creatine as it rephosphorylates ADP to ATP during muscular work and in post-mortem conditions. The creatine content of fish muscle varies depending on species ranging from 160 to 720 mg/100 g. White muscle tends to contain higher amounts of guanidine compounds than dark muscles. Invertebrates contain less creatine than finfish. There are other phosphogenes: arginine, glycocyamine, hypotaurocyamine, ophellin and lombricine. These compounds are the phosphorylated form of guanidine bases and are not present in the muscle of invertebrates.

2.1.4.5 Trimethylamine oxide (TMAO)

Trimethylamine oxide is a characteristic non-protein nitrogen compound in marine species. The amount of TMAO in the muscle varies according to species, age, size, season and environmental salinity. Demersal fish generally contain larger quantities of TMAO than pelagic fish, and the contents vary from 19 to 190 mg% (Venugopal 2009). Pelagic fish (sardines, tuna and mackerel) have their highest concentration of TMAO in the dark muscle while demersal fish have a much higher content in the white muscle. Elasmobranchs also contain high amounts of TMAO, while the content is small in molluscs and rather insignificant in freshwater fish species. There is a direct relationship between TMAO content and salinity of the habitat. TMAO seems to play a role in regulation of osmotic pressure in fish tissue and also protect the denaturation of protein. This compound is negligible in most freshwater fish (Venugopal 2006); however, some species like the Nile perch and tilapia contain TMAO.

The colourless, odourless and flavourless compound TMAO is degraded to trimethylamine (TMA) by bacterial spoilage and enzymatic, TMAO-reductase, activity. The species belonging to the family Enterobacteriaceae and some bacteria such as *Alteromonas*, *Photobacterium* and *Vibrio* are able to reduce TMAO due to being terminal electron donors (Stelo & Rehbein 2000). Formation of TMA depends primarily on the content of TMAO in the fish and gives the characteristic 'fishy' odour. The formation of dimethylamine (DMA) and formaldehyde from TMAO is due to the action of the indigenous enzyme TMAO demethylase. Generation of DMA and formaldehyde are correlated with textural change during frozen state. TMAO-breakdown products are measured to provide an indicator of fish freshness.

2.1.4.6 Urea

A high content of urea in fish muscle is characteristic for elasmobranchs such as sharks and rays. They are reported to produce, and retain, within their bodies large amounts of urea, a compound readily degraded to ammonia, leading to a rise in pH and total volatile basic nitrogen (TVB-N) during storage. The urea is broken down by the activity of bacterial urease with the formation of ammonia and carbon dioxide. In marine elasmobranchs, plasma osmolarity is higher than that of surrounding seawater, and osmoregulatory organic nitrogenous compounds, such as urea and TMAO, are high. Fresh water elasmobranchs retain and synthesize less urea than their marine counterparts.

2.1.4.7 Betaines

Glycine betaine is common in fish muscle. It plays a vital role in osmotic adjustment in various organisms and used as osmoprotectants in food systems. Betaines are abundant in molluscs and crustacean muscles, contributing to taste. Some marine fishes and invertebrates are reported to contain β-alanine betaine. Homorine is a metabolite of tryptophan and is common in invertebrates. It is widely accepted that homarine serves as an osmolyte in marine algae (Affeld *et al.* 2007).

2.2 Lipids

Lipids are found in all living organisms and play a role in the formation of the permeability barrier of cells, in the form of a lipid bilayer. Lipids are the major sources of cellular energy and function in living organisms where they are stored. The energy content per gram of lipid is 9.3 kcal, depending on the chain length. They also provide flavour, aroma, colour, texture, taste and nutritive value.

Lipids are the third major constituent in fish muscle after water and protein. The principal producers of marine lipids in the marine environment are microalgae. In fish muscle the lipids are triacylglycerol and phosphoglycerides, both containing long-chain fatty acids. The triglycerides are the source of stored energy and help conversion into phospholipids. The phospholipids (about 0.7–0.8% of the tissue) are the structural lipids containing small quantity of cholesterol. The phospholipids of tropical fish are more saturated than those in fish from temperate waters. Phytosterols are also present in bivalve molluscs, arising from microalgae and sediments. In elasmobranchs, such as sharks, the major quantity of lipid is stored in the liver and may consist of high-molecular-weight hydrocarbons like squalene.

The lipid content of fish species varies, even within the organs of species. These differences depend on many factors, such as the type of muscle and its location, age, sex and sexual maturation. Fish are often classified on the basis of their fat contents into lean (fat less than 5%), fat fish (fat 5–10%) and fatty fish (fat more than 10%) (Suriah *et al.* 1995). The distribution of lipid in fish muscle is heterogeneous, especially in fish with high lipid content. Lean fish store lipids only to a limited extent in the liver, whereas fatty fish store lipids in fat cells distributed in other body tissues located in the subcutaneous tissue, in the belly flap muscle, and in the muscles moving the fins and tail. The flesh of fatty fish is pigmented. Red muscles also contain two to five times more lipid, B-vitamins, glycogen and nucleic acids than white muscles. In many species fat content increases during the feeding season and its proportion decreases substantially after spawning. The lipid content is affected by external factors such as seasonal fluctuations in the environmental conditions and availability of phytoplankton. In many pelagic fish, lipid contents ranging from 12 to 20% are found during winter compared with 3–5% during summer (Venugopal 2009). It is known that there is an inverse relationship between unsaturated fatty acid content and environmental temperature for many marine fish.

Cholesterol is the main sterol in marine fish like haddock, pollock, salmon, and crustaceans like shrimp and lobster. It has been reported that fish muscle is low in cholesterol, wheras shrimps, prawns; squid and octopus are high in cholesterol. Freshwater fish muscle contains more cholesterol than marine fish. There were no significant correlations of cholesterol content of fish between the catching season and the catching ground (Oehlenschlager 2006).

Seafood lipids are known to provide high contents of important components for the human diet, such as nutritional lipid-soluble vitamins and essential and omega-3 poly-unsaturated fatty acids

(Ackman 1989). Marine-derived organisms differ from other sources, being the major sources of longer-chain and highly unsaturated fatty acids. The nature of fatty acids essentially determines the quality of lipids. Fatty acids rarely occur in the free form; they are mainly esterified to glycerolipids. The natural states of fatty acids are *cis* isomers, whereas processing may give rise to the formation of *trans* isomers. Fatty acids with carbon chains varying from 10 to 22 and unsaturation varying up to six double bonds are common. Depending on the nature of the hydrocarbonated chain, fish lipids are composed of saturated fatty acids (SFA), monounsaturated fatty acids (MUFA), and poly-unsaturated fatty acids (PUFA), whose proportions and amounts vary from one species to another.

2.2.1 Saturated fatty acids

SFA are long-chain carboxylic acids with no double bonds. Among SFA, palmitic and stearic acids are the important ones for marine-derived organisms. The fatty acid compositions of wild fish species such as bogue (*Boops boops*), mullet (*Mugil cephalus*), scad (*Trachurus mediterraneous*), sardine (*Sardinella aurita*), pandora (*Pagellus erythrinus*), red scorpion fish (*Scorpaena scrofa*), turbot (*Scopthalmus maeticus*) and common sole (*Solea solea*) ranged from 25.5% to 38.7% SFA, composed of mainly palmitic acid (15.5–20.2%) and stearic acid (3.32–7.27%) (Ozogul & Ozogul 2007). Lower SFA contents are reported for the fish caught from colder waters, in direct contrast to findings for fish caught from warmer temperatures (Huynh & Kitts 2009).

Palmitic acid (C16:0) was found the most abundant SFA in many algal species such as *Porphyra* sp., *Undaria pinnatifida, Laminaria* sp., *Hizikia fusiforme, Palmaria* sp, *Himanthalia elongata, Ulva lactuca* and *Durvillaea antarctica* (Sanchez-Machado *et al.* 2004; Ortiz *et al.* 2006; Dawczynski *et al.* 2007). Odd-number SFA pentadecanoic acid (15:0) and margaric acid (17:0) may be present in small amounts in some species such as bogue, seabream and seabass (Prato & Biandolino 2012). A high proportion of SFA, such as 20:0, has been observed in bivalves distributed in environments rich in organic material with an abundant bacterial load (Galap *et al.* 1999).

The recommended minimum value of PUFA/SFA ratio is 0.45 for a balanced diet (HMSO 1994; Simopoulos 2000). Many fish species have higher values than those recommended.

2.2.2 Mono-unsaturated fatty acids

MUFA contain one carbon double bond along the length of the hydrocarbon chain of the related fatty acid, while the other carbon atoms are linked with a single bond.

In the monounsaturated group, palmitoleic and oleic acids are the major constituents. Oleic acid is the dominating monounsaturated fatty acid in many freshwater fish species such as *Cyprinus carpio*, *Labeo rohita* and *Oreochromis mossambicus* (Jabeen & Chaudhry 2011). This fatty acid has exogenous origin and usually reflects the type of diet of the fish (Ackman 1989). In the muscle of many crustacean species such as *Nephrops norvegicus* (langoustine), *Palinurus vulgaris* (lobster) and *Penaeus kerathurus* (shrimp) C16:1n-7 and C18:1n-9 are the major MUFA (Tsape *et al.* 2010).

PUFA are the dominant group in the majority of species; however, there are some exceptions, for instance meagre and silver- and black-scabbard fish where the content of MUFA is higher than that of PUFA.

2.2.3 Poly-unsaturated fatty acids

Fatty acids with carbon chains varying from 10 to 22 and unsaturation varying from zero to six double bonds are common in marine-derived organisms. Approximately 50% of fatty acids in lean fish and 25% of fatty acids in fatty fish are PUFA. The group of PUFAs termed as omega-3 (n-3) and omega-6 (n-6) fatty acids have double bonds starting at three and six carbons from the methyl end of the fatty acid chain, respectively. The major varieties of PUFAs are n-3 PUFAs, such as docosahexaenoic acid (DHA, C22:6n-3) and eicosapentaenoic acid (EPA, C20:5n-3), and n-6 PUFAs such as arachidonic acid (AA, C20:4n-6). Other omega fatty acids such as linoleic (LA, C18:2n-6) and linolenic (ALA, C18:3n3) acids are present in fish lipids to a minor level.

The composition of fish lipids is different from that of other lipids because they are composed of mainly two types of fatty acids, EPA and DHA. These two main omega-3 fatty acids are typically found in marine fish and originate from the phytoplankton and seaweed that are part of their food chain. Seafood accumulates omega-3 fatty acids through phytoplankton, which are the primary producers of omega-3 fatty acids. Microalgae have the ability to synthesize and accumulate large amounts of omega-3 PUFA. Lipids are bioencapsulated by the algal cell wall (Patil *et al.* 2007). Zooplankton are the essential link between microalgae and pelagic marine environment (Berge & Barnathan, 2005).

Carbon fixed through photosynthesis is allocated to growth and cell division during the exponential growth phase of phytoplankton blooms. Therefore, the level of glycolipids is particularly high in this phase, and the proportion of n-3 PUFA can reach 50% of total lipids (Berge & Barnathan 2005). Molluscs having good access to phytoplankton accumulate a high proportion of EPA and DHA, which are reported to be essential for optimal growth of several juvenile bivalves. Comparing to marine fish, freshwater fish have lower DHA and EPA contents (Chedoloh *et al.* 2011). The difference can be attributed to the fact that fresh water fishes feed largely on vegetation and plant materials, whereas marine fish diets are mainly zooplanktons, rich in PUFA (Jabeen & Chaudhry 2011). Marine animals in the upper water layers gain nutrition through phytoplankton, which provides n-3 PUFA depending on solar energy. Shallow-water shrimp species provide a satisfying amount of PUFA. The levels of PUFAs of shallow-water shrimps (ranging from 33.44 to 42.77%) were found to be higher than those of deep-water shrimps (ranging from 29.68 to 33.95%) (Yerlikaya *et al.* 2013). Fat content and fatty acid composition of fish change according to life cycle, gender, food intake, location, temperature and salinity.

Arachidonic acid can be synthesized by humans from linoleic acid, and omega-3 fatty acids, as EPA and DHA, from linolenic acid; however, the conversion of these fatty acids is inadequate (Rubio-Rodriguez *et al.* 2010). Since the human body has a very limited capability to synthesize new omega-3 fatty acids, it is important to consume aquatic organisms to maintain healthy EPA and DHA levels.

It is well known that high level of PUFA in fish supply health benefits including reduced hypertension and cholesterol level, asthma, immune system disorders, susceptibility to mental illness, protection against heart disease, and improved brain and eye function in infants, development of nerve cells in growing children, and protection against cancer (Patil *et al.* 2007; Berra *et al.* 2009). FAO/World Health Organization (WHO) Expert Consultation on Fats and Oils in Human Nutrition (FAO/WHO 1994) recommends n-6/n-3 ratio of not more than 5.0. Consumption of foods rich in n-3 PUFAs will lead to keep n-6/n-3 fatty acid ratio less than 4 and the PUFA/SFA ratio will be higher than 0.4 (Wood *et al.* 2003). It is essential to decrease n-6 intake while increasing n-3 in the prevention of chronic disease (Simopoulos 2002). The ratio of n-6/n-3 of the edible tissue of shrimp *Metapenaeus monoceros* was 0.795, whereas this value was 0.152 in *Aristeomorpha foliacea*. The ratio of n-6/n-3 is lower in marine fish species. These values for sardine, anchovy and picarel were 0.079, 0.114 and 0.090%, respectively (Zlatanos & Laskaridis 2007). Dietary recommendations suggest that the consumption of omega-3 should be increased and minimal average daily intake of omega-3 (EPA+DHA) is 0.2 g per person (Kolanowski *et al.* 2007). Sufficient intake of EPA and DHA is vital in maintaining a healthy life. EPA+DHA levels ranged from 23.45% in *Penaeus semisulcatus* to 29.06% in *Penaeus kerathurus* (Yerlikaya *et al.* 2013). EPA+DHA values ranged between 23 and 32% in wild marine shrimp (Bragagnolo and Rodriguez-Amaya 2001). This value was 33.35% in sardine, 26.69% in anchovy and 29.08% in picarel caught in December (Zlatanos & Laskaridis 2007). People are encouraged to consume fish and other seafood.

There is an increasing worldwide interest in the use of fish and fish oil. The researchers are dealing with fish oil-enriched products such as milk, yoghurt, drinking yoghurt, margarine, mayonnaise and emulsions (Sorensen *et al.* 1998; Timm-Heinrich *et al.* 2004; Kolanowski *et al.* 2007; Gökoğlu et al. 2012). The major experienced problem in fish oil-enriched products is sensitivity of PUFA to oxidative degradation and herewith strong fishy odour. PUFA are highly susceptible to oxidation on exposure to air, even at ambient temperature, leading to oxidative rancidity (Aubourg 2010). Enzymes such as lipoxygenase, peroxidase and microsomal enzymes from animal tissues can potentially initiate lipid peroxidation producing hydroperoxides. The other type of problem is the development of hydrolytic rancidity leading to the formation of free fatty acids. These free fatty acids react with proteins causing protein denaturation. Lipid hydrolysis is more in ungutted than in gutted fish, probably due to the involvement of lipases present in digestive enzymes.

2.3 Carbohydrates

The carbohydrates are a major source of metabolic energy in the living organism. Carbohydrates occur in glycogen and as part of the chemical constituents of nucleotides in fish muscle. The amount of glycogen content is associated with the ultimate pH value of the flesh. This value is important, not only for texture, but also for other factors, such as water holding-capacity, bacterial growth and colour.

The content of carbohydrates is influenced by conditions before and during capture, which may lead to depletion of glycogen stores. The amount of lactic acid produced is related to the amount of stored glycogen in the living tissue. In general, fish muscle contains low level of glycogen compared to mammals; therefore less lactic acid is generated after death. The fish exposed to various forms of long-term stress before death, which depletes its glycogen reserves and consequently on the ultimate post- mortem pH.

The carbohydrate content in marine derived organisms is low and practically considered zero (Payne *et al.* 1999). Whereas, it was reported that some marine invertebrates are characterized by a high content of carbohydrates; up to 10.2% and 12.5% total sugars can be found in subcuticular tissue of spiny lobster and blue crab, respectively, with the highest amounts of glucose followed by galactose and mannose. Seasonal variations of glycogen content in mussels (*Mytilus edulis*) are also high, showing values in the range 4%–37% of tissue dry weight. (Falch *et al.* 2010). Oyster contained significantly high carbohydrate at 6.45 % (Nurnadia *et al.* 2011).

2.4 Minerals

Vitamins and minerals are essential for life since their deficiencies lead to various health disorders. Minerals play important roles in the enzymatic reactions, formation of the skeletal structure, muscle contraction, meintanance of colloidal systems (osmotic pressure, viscosity, diffusion) and regulation of acid–base balance. Minerals are divided into two groups: macro- and microelements. Macroelements are those present in amounts greater than 5 g in the human body and include calcium, phosphorus, potassium, sulphur, sodium, chloride and magnesium. Microelements are present in microgram levels per 1 g of food. They are both essential for human health.

All kinds of fish and shellfish present a well-balanced content of most minerals. Mineral compounds are taken up and accumulated by aquatic organisms both from the surrounding medium and via food sources. The mineral content varies depending on season, biological characteristics (species, age, size, gender, sexual maturity), food sources, environmental factors (temperature and salinity of the surrounding water, presence of planktons, contaminants) and processing.

Marine-derived organisms are particularly valuable sources of minerals such as calcium and phosphorus, as well as iron, copper and selenium. The total mineral content in wet fish meat ranges from 0.6 to 1.5%. Calcium and phosphorus account for more than 75% of the mineral in the skeleton. The skeleton of the fish contains 35–58% protein and up to 65% inorganic materials, depending on the age. In general, shellfish tend to be richer sources of minerals than fish. An average serving of fish or marine invertebrate can satisfy the total human requirements for essential microelements.

2.4.1 Macroelements

Calcium and phosphorus are the most abundant minerals in fish, human and other organisms, and they are mostly present in the bones. Calcium is needed to form bones and regulatory functions. Phosphorus is a major component of bone, teeth and is a regulator of energy metabolism. Fishery products are good sources of these minerals. The bones of small fish are frequently eaten with the flesh, which increases the intake of calcium and phosphorus. Oysters, clams and shrimp contain more calcium than other fish and meat because of shell formation and muscle function. Products derived from bones and other calciferous tissues of seafood processing wastes are useful because of their high mineral content. Seaweed species also contain calcium and phosphorus.

Phosphorus is available as salts of calcium, magnesium and sodium. Magnesium is found in bones followed by muscles, soft tissues and body fluids. It is essential for energy metabolism and synthesis of proteins. Fishery products are poor sources of magnesium. Fish red blood cells contain magnesium.

Potassium, sodium and chloride are electrolytes that carry out osmoregulation, regulation of acid–base balance and transport across the cell membrane. Potassium promotes cellular growth and helps maintain normal blood pressure, while chlorine allows the maintenance of electrolyte balance. Fresh seafood is low in sodium, which makes it suitable for low-sodium diets. However, the content of this mineral increases during processing by the addition of salt or sodium-containing compounds. There is a wide variation in the sodium content of fishes. The sodium content is 60 mg/100 g in freshwater and marine fish, and 120–140 mg/100 g in shellfish. The potassium concentration is reported to be higher than the sodium content, ranging from 198 to 440 mg/100 g in seafood (Gökoğlu 2002). Also, crabs are rich in terms of mineral content, especially sodium, potassium, calcium and phosphorus (Gökoğlu & Yerlikaya 2003).

2.4.2 Microelements

Consumption of an average portion of fish and marine invertebrates meets the daily needs in terms of essential microelements.

Iron is a crucial component of oxygen-carrying proteins, haemoglobin, myoglobin and cytochromes. Iron is also involved in the action of many enzymes as a cofactor. In meat products, heme and non-heme irons are found in a ratio of 2:3. Heme iron found in meat is more absorbable than non-heme iron found in plants and vegetables. Dark muscles of fish tend to contain more iron than white muscles. However, it is reported that the iron content of beef muscle is aproximately three times higher than fish muscle.

Zinc promotes immune system, growth and repair of tissue. Zinc is present in over 70 enzymes, allowing regulation of many metabolic activities. Oysters are the richest zinc source of animal origin, whereas the lowest concentration of zinc is present in fish and mammals, among marine organisms. Liver, viscera, gills, scales and milt contain higher concentrations of zinc than muscles of fish. Zinc toxicity can cause copper deficiency and can harm the immune system.

Molluscs and crustaceans are good sources of zinc and copper, as well as iodine. Copper, which is necessary for maintenance of blood vessels, tendons and bone, is sufficiently available from crustaceans, especially lobsters. Fish muscle contains as much copper as land animals. The accumulation of copper in fish is affected by season, temperature, salinity and the presence of other metals such as manganese and iron in water.

Manganese is available from crustaceans in sufficient quantities, especially lobsters, and can satisfy the total human requirement of essential micro elements. Molluscs and crustaceans contain significantly higher levels of manganese than fish.

Iodine and fluorides are also present in relative abundance in fish. Marine fish and shellfish are rich sources of iodine, being highest in oysters, followed by clams, lobster, shrimp, crayfish and ocean fish. The iodine content of freshwater fish is lower than marine fish. Saltwater fish is an excellent source of iodine. Fish has the highest fluorine content among animal-originated food sources (Toth & Sugar 1978).

All marine-derived organisms are important sources of selenium and iodine. Fish and shellfish contain much higher concentrations of selenium than other meats. Selenium provides protection against mercury and cadmium toxicity. Certain fishes like tuna are especially good sources of macro minerals like magnesium and trace elements like selenium.

Some microelements – toxic heavy metals such as mercury, arsenic, lead and cadmium – do not need to be in high concentrations in order to cause harm. Heavy metals are recognized as one of the most important pollutants, and their accumulations in organisms are monitored for the safety of seafood consumption. Fish take up heavy metals from their food and the water they pass through their gills. The areas of sea where there is high concentration of phytoplankton, the marine derived organism tend to accumulate heavy metals more than other areas. Some species living long lives, especially predators, are known for storing higher amount of heavy metals in different organs (Oehlenschlager 2002). Invertebrates tend to accumulate more metals than fish as a result of differences in the evolutionary strategies adopted by various phyla. Crustaceans are reported to be important bioindicators of heavy metal pollution (Gökoğlu et al. 2008).

2.5 Vitamins

Vitamins are organic compounds that are essential for metabolic reactions in the body. They function as cofactors or coenzymes in biochemical reactions. Vitamins are not the source of energy; however, they enable the assimilation of carbohydrates, proteins and fats. They also play a critical role in the formation of blood cells, hormones and neurotransmitters. Vitamins are present and are effective in minute amounts; most of them cannot be synthesized by the organism and their absence from the diet causes specific deficiencies. Therefore, they are essential nutrient elements that shoud be supplied in the diet. Vitamins are divided into two groups basing on their solubility in fat (A, D, E, K) and water (B complex and C). Both water-soluble and fat-soluble vitamins are present in fish.

The amount of vitamins and minerals is species specific and can vary with many factors such as season, age, feeding and environmental factors. Vitamins are found in the body or liver depending on whether the fish is lean or fat. Fatty species supply reasonable amounts of vitamins A and D, which are found especially in fish liver oils. These vitamins continue to accumulate in the liver with increasing age. Some mobilization of vitamins from muscle to other tissues may occur during sexual maturity. Water-soluble vitamin composition of wild fish may alterate during migration, maturation, and in when food is scarce (Lall & Parazo 1995).

2.5.1 Fat-soluble vitamins

Fat-soluble vitamins A, D and E are present in seafood in varying amounts, often in concentrations higher than those in other meats.

Vitamin A (retinol) as retinoids, primarily retinyl esters, is abundant in some marine derived foods, whereas carotenoids are responsible for the colour of fish and shellfish. The human body converts β-carotene in the diet into vitamin A. Carotenoids are highly concentrated in fish liver oils, but small amounts are found in fish muscles or fillets. The concentration of vitamin A in the liver depends on the species and other factors such as fish size, spawning cycle, season and feeding. There is one more –CH=CH– compound present in the structure of vitamin A in freshwater fish than marine fish, defining vitamin A_2. Exposure to air and heat, and storage time also influence the destruction of vitamin A compounds.

The meat of fatty or semi-fatty fishes is an excellent source of vitamin D. Like in humans, vitamin D is synthesized from 7-dehydrocholesterol in the fish after exposure to ulrtaviolet (UV) rays from the sun. UV is absorbed by the water. Thus, much of vitamin D in fish is dietary origin derived from plankton, acting as a source of both vitamin D_2 and D_3 (Lund 2013). This vitamin is known for its antirachitic activity. Fish liver oils are the richest sources of vitamin D, and other marine-derived organisms such as finfish and shellfish are natural vitamin D contributors. The concentration of vitamin D is lower in elasmobranches than teleost fishes due to the less vitamin D requirement of elasmobranchs for the calcification of their structural tissue (Gökoğlu 2002). Food processing, cooking and storage of foods do not generally affect the concentration of vitamin D.

Seafood provides small amounts of vitamin E. Vitamin E exists in eight different forms: four tocopherols (α-, β-, γ- and δ-tocopherols) and four tocotrienols (α-, β-, γ- and δ-tocotrienols). α-Tocopherol, a natural antioxidant, has been found to be the principal tocopherol in marine animals that protects fatty acids from oxidative degradation. Fish cannot synthesize vitamin E, and hence, the concentration of this vitamin is related to feed. The dark muscle of fish has more vitamin E than white muscle, and shellfish has little vitamin E content. Krill is a very good source of vitamin E (Kim 2010).

The antihaemorrhagic factor, vitamin K, is also present in fish. The abundant presence of fat-soluble vitamins in the liver of lean fish drew attention of the industry as vitamin supplements. The marine mammalians and fish such as cod, mackerel, shark and swordfish are regarded as potential vitamin sources. Liver oils from shark and tuna are rich in vitamin A.

2.5.2 Water-soluble vitamins

Fish are considered a good source of B-complex vitamins. Fish liver, eggs, milt and skin are good sources of thiamine (B_1), riboflavin (B_2), B_6, pantothenic acid, biotin and B_{12}.

Seafood provides moderate amounts of thiamine (B_1). Most fish and seafood have small amounts of thiamine in the form of thiamine mono-, di-, triphosphates, varying with species. The content of thiamine varies individually within the same species depending on the metabolic demand. Redundant water-soluble vitamins tend to be discarded from the body. Dark-coloured fish meat contains more thiamine than the others, generally concentrated in the liver and ovary. Raw fish and shellfish contain thiaminase, an enzyme that destroys thiamine, which is inactivated during cooking process. Thiamine is destroyed by heat and oxygen, or is lost in cooking water or when exposed to low-dose radiation.

Riboflavin (B_2) acts with the protein-bound coenzymes, flavin adenine dinucleotide (FAD) and flavin mononucleotide (FMN), and plays a role in conversion of energy from fats and carbohydrates. Seafood is generally a modest source of riboflavin, which is less than in other meat products. Some species like mackerel and squid are good sources. Metabolically active tissues of fish are rich in riboflavin. The dark muscles contain 10–20 times more riboflavin than the white muscles. Riboflavin is stable to heat and acidic conditionsbut it is oxidized if light is excluded.

Niacin (B_3) contents in fish and seafood vary depending on the variety of fish or seafood. Lean white fish and shellfish tend to contain smaller amounts of niacin than fatty fish species such as mackerel, salmon and tuna. Fish and shellfish are good sources of vitamin B_6, which is especially found in liver, ovary and spawn. Pelagic species are rich in this vitamin. Freshwater fish tend to contain lower vitamin B_6 than marine species. Vitamin B_6 is unstable to visible and UV light, but is stable in acidic conditions. Fish and seafood only contain a moderate amount of pantothenic acid (B_5) except for salmon, trout and abalone, which are good B_5 sources. The vitamin content of cultured fish flesh changes depending on feeding. This vitamin is located in ovaries, dark muscles and heart. Fish and shellfish, especially oysters, are rich sources of vitamin B_{12}. Dark muscles of fish tend to contain high B_{12} amounts compared to white muscles. Unlike other water-soluble vitamins, B_{12} is stored in liver (Lall & Parazo 1995). Large losses of water-soluble vitamins can occur in the blanching or boiling of foods, leaching into cooking water, and by drip-loss during freeze–thaw cycle.

Muscles of wild and cultivated marine and freshwater fish generally contain low amounts of vitamin C (ascorbic acid), usually not exceeding 1 mg/100 g. The loss of ascorbic acid in fish and fishery products occurs during storage and processing, due to its sensitivity to oxidation. Liver, kidney and brain contain high concentrations of ascorbic acid.

2.6 Conclusion

Knowledge of the chemical composition of fish is important in order to find out nutritional properties, their usage, storage stabilities and the applied processing technology. The content of proximate composition, amino acids, fatty acids, minerals and vitamins of some selected fish species are given in Tables 2.1, 2.2, 2.3, 2.4 and 2.5, respectively. The

Table 2.1 Proximate composition of selected fish species*

Parameter (%)	Sardine	Mackerel	Rainbow trout	European eel	Wild meagre	Cultured meagre	Wild seabass	Cultured seabass
Moisture	71.13	76.37	73.38	60.4	76.36	75.02	69.46	72.63
Protein	18.10	19.87	19.80	16.3	22.10	20.40	17.64	23.37
Lipid	9.03	1.96	3.44	21.8	1.23	3.12	9.19	6.66
Ash	1.03	1.14	1.35	1.3				

*Luzzana et al. (2003); Gökoğlu et al. (2004b); Periago et al. (2005); Cakli et al. (2006); Fernandes et al. (2014). Meagre, genus Argyrosomus.

Table 2.2 Amino acid contents of selected fish species*

Amino acids (g/100 g)	Atlantic salmon	Channel catfish	Rainbow trout	Horse mackerel	Yellow perch	Scabbard fish	Mackerel	Herring
Threonine	4.95	4.41	4.76	4.28	3.19	4.46	4.24	4.55
Valine	5.09	5.15	5.09	4.53	4.12	5.73	5.41	4.21
Methionine	1.83	2.92	2.88	2.30	2.46	3.18	2.48	2.53
Isoleucine	4.41	4.29	4.37	4.57	3.43	5.10	4.77	4.22
Leucine	7.72	7.40	7.59	7.21	6.82	8.28	6.86	7.03
Phenylalanine	4.36	4.14	4.38	4.55	3.38	3.82	4.41	4.63
Lysine	9.28	8.51	8.49	6.82	6.06	10.19	8.02	7.05
Histidine	3.02	2.17	2.96	2.87	2.07	1.91	3.41	2.90
Arginine	6.61	6.67	6.41	5.31	5.03	6.37	7.07	5.13
Serine	4.61	4.89	4.66	4.52	3.31	3.82	4.83	4.20
Aspartic acid	9.92	9.74	9.94	9.61	8.13	10.19	9.05	9.60
Glutamic acid	14.31	14.39	14.22	12.02	12.23	15.29	13.91	13.36
Glycine	7.41	8.14	7.76	5.32	3.20	5.10	5.49	5.97
Alanine	6.52	6.31	6.57	5.51	4.37	6.37	5.22	5.19
Tyrosine	3.50	3.28	3.38	2.86	2.97	3.82	3.42	2.72
Proline	4.64	6.02	4.89	5.59	3.10	5.10	5.13	7.33
Cysteine	0.80	1.34	0.86					
Tryptophan	0.93	0.78	0.93	0.95	1.06		1.02	0.91

*Mai *et al.* (1980); Bandarra *et al.* (2009); Sharma & Katz (2013); Oluwaniyi *et al.* (2010).

Table 2.3 Fatty acid composition of selected fish species*

Fatty acids (%)	Sardine	Mackerel	Seabass	Red mullet	Picarel	Scabbard fish	Anchovy	Sprat	Black sea goby
12:0	0.23	1.18	0.64	0	0.12				3.83
14:0	4.28	1.38	4.46	3.34	4.71	2.12	5.05	4.00	4.88
15:0	0.08	3.06	0.00	0.63	1.14	0.38			
16:0	20.38	16.37	29.92	29.17	30.44	12.20	20.0	24.3	10.88
17:0	0.40	1.22	0.51	0.24	0.42	0.34			2.77
18:0	7.91	6.77	6.58	6.57	4.87	4.24	5.87	3.49	5.80
ΣSFA	33.26	30.29	42.93	39.95	42.27	19.36	33.2	33.2	43.96
16:1n-7	7.45	2.72	4.96	9.53	8.71	3.26	5.46	5.67	3.40
17:1	1.07	0.32	0.54	0.00	0.54	0.24			2.77
18:1n-9	4.26	10.10	10.86	13.87	9.47	26.06	5.40	16.1	6.62
18:1n-7						6.28	3.29	2.40	
20:1n-9	0.86	0.59	1.42	1.89	1.57	14.10			3.06
24:1n-9			1.05	0.27	0.83	0.19			2.70
ΣMUFA	14.10	13.96	24.62	32.39	25.61	65.67	15.1	28.1	25.83
18:2n-6	1.84	3.35	3.77	1.42	1.33	1.09	1.62	1.03	1.26
18:3n-3	0.67	1.56	1.44	1.24	2.07		1.29	0.99	4.16
18:4n-3						0.11	1.71	2.00	
20:2n-6	0.61	0.35	0.00	0.17	0.73	0.14			1.77

20:4n-6	1.04	1.34	2.80	3.01	1.76		2.10	0.58	4.68
20:5n-3	10.66	7.51	6.91	7.16	7.21	1.94	12.7	7.90	3.04
22:5n-3	2.98	1.87				0.84	1.04	0.46	
22:6n-3	30.20	27.54	13.83	12.05	14.67	5.65	19.1	17.6	8.88
ΣPUFA	**48.81**	**35.66**	**32.45**	**27.66**	**32.12**	**9.51**	**43.3**	**32.7**	**30.20**
Σn-3	45.31	38.62	23.12	20.81	25.22	8.54	36.7	29.6	13.17
Σn-6	3.50	4.90	9.33	6.39	5.73	0.97	5.69	2.28	4.99
n-3/n-6	12.94	7.88	2.47	3.25	4.40	8.80	6.44	12.98	2.63

*Bandarra *et al.* (2009); Pirini *et al.* (2010); Stancheva *et al.* (2010); Prato & Biandolino (2012); Fernandes *et al.* (2014).

Table 2.4 Mineral content of selected fish species**

Minerals (mg/100g)	Horse mackerel	Rainbow trout	Sardine	Hake	African catfish	Blue whiting	Cuttlefish	Spanish mackerel
Potassium (K)	403	306	404	320	181	388	3.13	13.53
Phosphorus (P)	263	337	296	421		604		
Sodium (Na)	8C	45	65	143	31	136	86.24	36.20
Magnes um (Mg)	33	41	29	36.9	18	36.7	1317	874.7
Calcium (Ca)	69	63	70	25.6	4.01	17.7	19.43	19.7
Iron (Fe)	1.2	0.21	1.7	0.51	1.2	0.40	252.07*	319.89*
Zinc (Zr)	1.2	0.96	1.7	0.41	0.34	0.53	413.83*	227.27*
Manganese (Mn)	<0.02	0.07	<0.02		0.03		11.93*	9.31*
Copper (Cu)	0.08	0.03	<0.03	0.07	0.21	0.29	163.93*	60.50*

*µg/100 g wet weight

**Martinez-Valverde et al. (2000); Gökoğlu et al. (2004b); Nunes et al. (2006); Ersoy & Ozeren (2009); Nurnadia et al. (2013).

Table 2.5 Vitamin contents of selected fish species*

Vitamins	Chub mackerel	Common sole	Hake	Atlantic cod	Atlantic salmon	European eel	Monkfish	Tuna	African catfish
Vit A µg	28	4.4	7.3	3.8	33	887	24	11	18.1
Vit E mg	1.3	0.32	0.51	0.3	4	2.4	0.23	0.64	0.34
Vit D µg	2.4	9.4	1.1	4.5	11	16	0.00	4.2	
Vit B_1 mg	0.13	0.09	0.04	0.05	0.18	0.28	0.04	0.09	0.07
Vit B_2 mg	0.23	0.13	0.03	0.07	0.04	0.26	0.02	0.04	0.03
Vit B_6 mg	1.0	0.33	0.05	0.07	0.45	0.15	0.05	0.56	0.08
Niacine mg	9.0	2.8	0.74	0.8	0.00	1.3	2.0	10	1.13
Folate µg	14	10	13	8.1	10	9.3	7.3	8.3	
Vit B_{12} µg	14	0.94	0.47	0.95	0.00	0.00	0.00	2.4	

*Dias *et al.* (2003); Nunes *et al.* (2006); Ersoy & Ozeren (2009).

chemical composition of fish varies depending on age, season, feeding cycle, migratory swimming, sexual maturation and environment (Rasmussen 2001). The factors affecting fish raised in aquaculture are more controlled. Therefore, the composition of fish can be estimated considering feed composition, environment, fish size and genetic characteristics.

References

Ackman, R. G. (1989). Fatty acids. In R. G. Ackman (Ed.), *Marine Biogenic Lipids, Fats and Oils*, pp. 145–178, CRC Press, Boca Raton.

Affeld, S., Wägele, H., Avila, C., Kehraus, S. & Konig, G.M. (2007). Distribution of homarine in some Opisthobranchia (Gastropoda: Mollusca). *Bonner Zoologische Beiträge*, **55**: 181–190.

Auborg, S.P. (2010). Lipid Compounds, In: *Handbook of Seafood and Seafood Products Analysis* (Eds. Nollet, L.M.L. & Toldra, F.), pp. 70–82, CRC Press, Boca Raton.

Bandarra, N.M. Batista, I. & Nunes, M.L. (2009). Chemical composition and nutritional value of raw and cooked black scabbardfish (*Aphanopus carbo*). *Scientia Marina*, 105–113.

Berra, B., Montorfano, G., Negroni, M., Corsetto, P. & Rizzo, A.M. (2009). Biomarkers of long-chain PUFA omega-3 fatty acids and the human nutritional status. *Lipid Technology*, **21** (2): 32–35.

Berge, J. & Barnathan, G. (2005). Fatty acids from lipids of marine organisms: molecular biodiversity, roles as biomarkers, biologically active compounds, and economical aspects. *Advances in Biochemical Engineering/Biotechnology*, **96**: 49–125.

Bragagnolo, N. & Rodriguez-Amaya, D. B. (2001). Total lipid, cholesterol, and fatty acids of farmed freshwater prawn (*Macrobrachium rosenbergii*) and wild marine shrimp (*Penaeus brasiliensis, Penaeus schimitti, Xiphopenaeus kroyeri*). *Journal of Food Composition and Analysis*, **14**: 359–369.

Cakli, S., Dincer, T., Cadun, A., Saka, S. & Firat, K. (2006). Seasonal variation of proximate and fatty acid class composition of wild and cultured Brown meagre (*Sciena umbra*), a new species for aquaculture. In: *Seafood Research from Fish to Dish*, (Eds. Luten, J.B., Jacobsen, C., Bekaert, K., Saebo, A. & Oehlenschlager, J.), pp. 469–476. Wageningen Academic Publishers, Netherlands.

Chan, J.K., Gill, T.A. & Paulson, A.T. (1992). Cross-linking of myosin heavy chains from cod, herring and silver hake during thermal setting. *Journal of Food Science*, **7**: 906–12.

Chedoloh, R., Karilla, T.T. & Pakdeechanuan, P. (2011). Fatty acid composition of important aquatic animals in Southern Thailand. *International Food Research Journal*, **18**: 758–765.

Dawczynski, C., Schubert, R. & Jahreis, G. (2007). Amino acids, fatty acids, and dietary fibre in edible seaweed products. *Food Chemistry*, **103**: 891–899.

Dias, M.G., Sanchez, M.V., Bartolo, H. & Oliviera, L. (2003). Vitamin content of fish and fish products consumed in Portugal. *Electronic Journal of Environmental, Agricultural and Food Chemistry*, **2** (4): 510–513.

Ersoy, B. & Ozeren, A. (2009). The effect of cooking methods on mineral and vitamin contents of African catfish. *Food Chemistry*, **115**: 419–422.

Falch, E., Overrein, I., Solberg, C. & Slizyte, R. (2010). Composition and Calories. In: *Handbook of Seafood and Seafood Products Analysis* (Eds. Nollet, L.M.L. & Toldra, F.), pp. 258–276, CRC Press, Boca Raton.

FAO/WHO (1994). Fats and oils in human nutrition. *Report of a Joint FAO/WHO Expert Consultation*, 19 to 26 October 1993, Rome. 168 pp.

Fernandes, C.E., Vasconcelos, M.A.S., Ribeiro, M.A., Sarubbo, L.A., Andrade, S.A.C. & Filho, A.B.M. (2014). Nutritional and lipid profiles in marine fish species from Brazil. *Food Chemistry*, **160**: 67–71.

Galap, C., Netchitailo, Leboulenger, F. & Grillot, J. (1999). Variations of fatty acid contents in selected tissues of the female dog cockle (*Glycymeris glycymeris* L., Mollusca, Bivalvia) during the annual cycle. *Comparative Biochemistry and Physiology Part A: Molecular & Integrative Physiology*, **122** (2): 241–254.

Gill, T. (2000). Nucleotide-degrading enzymes. In: *Seafood Enzymes* (Eds. Haard, N.F. & Simpson, B.K.), pp. 37–68, Marcel Dekker Inc., New York.

Gökoğlu, N. (2002). *Su Ürünleri İşleme Teknolojisi.* pp. 157, Su Vakfı Yayınları, Istanbul.

Gökoğlu, N. & Yerlikaya, P. (2003). Determination of proximate composition and mineral contents of blue crab (*Callinectes sapidus*) and swim crab (*Portunus pelagicus*) caught off the gulf of Antalya. *Food Chemistry*, **80**: 495–498.

Gökoğlu, N., Yerlikaya, P. & Cengiz, E. (2004a). Changes in biogenic amine contents and sensory quality of sardine (*Sardina pilchardus*) stored at 4°C and 20°C. *Journal of Food Quality*, **27** (3): 221–231.

Gökoğlu, N., Yerlikaya, P. & Cengiz, E. (2004b). Effects of cooking methods on the proximate composition and mineral contents of rainbow trout (*Oncorhynchus mykiss*). *Food Chemistry* **84**: 19–22.

Gökoğlu, N., Gökoğlu, M. & Yerlikaya, P. (2008). Trace elements in edible tissues of three shrimp species (*Penaeus semisulcatus, Parapenaeus longirostris, Paleomon serratus*). *Journal of the Science and Food and Agriculture*, **88**: 175–178.

Gökoğlu, N., Topuz, O.K., Buyukbenli, H.A. & Yerlikaya, P. (2012). Inhibition of lipid oxidation in anchovy oil (*Engraulis encrasicholus*) enriched emulsions during refrigerated storage. *International Journal of Food Science and Technology*, **47**: 1398–1403.

Haard, N.F. (1995). Composition ond nutritive value of fish proteins and other nitrogen compounds. In: *Fish and Fishery Products*, pp. 77–115, CAB International Press, Wallingford, UK.

Hargin, K.D. (2002). Measurement of the fish content in fish products. In: *Seafoods Quality, Technology and Nutraceutical Applications*, pp. 58–71, Springer, Berlin.

Harnedy, P.A. & FitzGerald, R.J. (2012). Bioactive peptides from marine processing waste and shellfish: A review. *Journal of Functional Foods*, **4**: 6–24.

Hedges N. (2002). Maintaining the quality of frozen fish. *Safety and Quality Issues in Fish Processing* (Ed Bremner, H.A.), pp. 379–400, Woodhead Publishing, Cambridge, UK.

HMSO, UK. (1994). *Nutritional aspects of cardiovascular disease* (report on health and social subjects No. 46), HMSO, London.

Huss, H. H. (1995). Quality and quality changes in fresh fish. FAO Fisheries Technical Paper 348. [Online], Available from http://www.fao.org/3/a-v7180e/index.html (accessed 9 January 2015).

Huynh, M.D. & Kitts, D.D. (2009). Evaluating nutritional quality of pacific fish species from fatty acid signatures. *Food Chemistry*, **114**: 912–918.

Jabeen, F. & Chaudhry, A.S. (2011). Chemical compositions and fatty acid profiles of three freshwater fish species. *Food Chemistry*, **125**: 991–996.

Kikuchi, K., Matahira, Y. & Sakai, K. (2004). Separation and physiological functions of anserine from fish extract. *Developments in Food Science*, **42**: 97–105.

Kim, Y. (2010). Vitamins. *Handbook of Seafood and Seafood Products Analysis*, pp. 327–350, CRC Press, Boca Raton.

Kolanowski, W., Jaworska, D. & Weissbrodt, J. (2007). Importance of instrumental and sensory analysis in the assessment of oxidative deterioration of omega-3 long-chain polyunsaturated fatty acid-rich foods. *Journal of the Science of Food and Agriculture*, **87**: 181–191.

Lall, S.P. & Parazo, M.P. (1995). Vitamins in fish and shellfish. *Fish and Fishery Products*, pp. 77–115, CAB International Press.

Lund, E.K. (2013). Health benefits of seafood; is it just the fatty acids? *Food Chemistry*, **140**: 413–420.

Luzzana, U., Scolari, M., Campo Dall'Orto, B., *et al.* (2003). Growth and product quality of European eel (*Anguilla anguilla*) as affected by dietary protein and lipid sources. *Journal of Applied Ichthyology*, **19**: 74–78.

Mai, J., Shetty, J.K., Kan, T. & Kinsella, J.E. (1980). Protein and amino acid composition of select freshwater fish. *Journal of Agriculture and Food Chemistry*, **28**: 884–885.

Martinez-Valverde, I., Periago, M.J., Santaella, M. & Ros, G. (2000). The content and nutritional significance of minerals on fish flesh in the presence and absence of bone. *Food Chemistry*, **71**: 503–509.

Nunes, M.L., Bandarra, N., Oliveira, L., Batista, I. & Calhau, M.A. (2006). Composition and nutritional value of fisheery products consumed in Portugal. In: *Seafood Research from Fish to Dish*, pp. 477–487, Wageningen Academic Publishers, Netherlands.

Nurnadia, A.A., Azrina, A. & Amin, I. (2011). Proximate composition and energetic value of selected marine fish and shellfish from the West coast of Peninsular Malaysia. *International Food Research Journal*, **18**: 137–148.

Nurnadia, A.A., Azrina, A., Amin, I., Mohd Yunus, A.S. & Mohd Izuan Effendi, H. (2013). Mineral contents of selected marine fish and shellfish from the west coast of Peninsular Malaysia. *International Food Research Journal*, **20** (1): 431–437.

Oehlenschlager, J. (2002). Identifying allergens in fish. In: *Safety and Quality Issues in Fish Processing*, pp. 95–113, Woodhead Publishing, Cambridge, UK.

Oehlenschlager, J. (2006). Cholesterol content in seafood, data from the last decade: A review. *Seafood Research from Fish to Dish*, pp. 41–57, Wageningen Academic Publishers, Netherlands.

Olafsdottir, G. & Jonsdottir, R. (2010). Volatile aroma compounds in fish. In: *Handbook of Seafood and Seafood Products Analysis* (Eds. Nollet, L.M.L. & Toldra, F.), pp. 97–113, CRC Press, Boca Raton.

Ortiz, J., Romero, N., Robert, P., *et al.* (2006). Dietary fiber, amino acid, fatty acid and tocopherol contents of the edible seaweeds *Ulva lactuca* and *Durvillaea antarctica*. *Food Chemistry*, **99**: 98–104.

Oluwaniyi, O.O., Dosumu, O.O. & Awolola, G.V. (2010). Effect of local processing methods (boiling, frying and roasting) on the amino acid composition of four marine fishes commonly consumed in Nigeria. *Food Chemistry*, **123**: 1000–1006.

Ozogul Y. & Ozogul, F. (2007). Fatty acid profiles of commercially important fish species from the Mediterranean, Aegean and Black Seas. *Food Chemistry*, **100**: 1634–1638.

Patil, V., Kallqvist, T., Olsen, E., Vogt, G. & Gislerod, H.R. (2007). Fatty acid composition of 12 microalgae for possible use in aquaculture feed. *Aquaculture International*, **15**: 1–9.

Payne, S.A., Johnson, B.A. & Otto, R.S. (1999). Proximate composition of some north-eastern Pacific forage fish species. *Fish Oceanography*, **8** (3): 159–177.

Periago, M.J., Ayala, M.D., Lopez-Albors, O., *et al.* (2005). Muscle cellularity and flesh quality of wild and farmed sea bass, *Dicentrarchus labrax* L. *Aquaculture*, **249**: 175–188.

Pirini, M., Testi, S., Ventrella, V., Pagliarani, A. & Badiani, A. (2010). Blue-back fish: Fatty acid profile in selected seasons and retention upon baking. *Food Chemistry*, **123**: 306–314.

Prato, E. & Biandolino, F. (2012). Total lipid content and fatty acid composition of commercially important fish species from the Mediterranean, Mar Grande Sea. *Food Chemistry*, **131**: 1233–1239.

Rasmussen, R.S. (2001). Quality of farmed salmonids with emphasis on proximate composition, yield and sensory characteristics. *Aquaculture Research*, **32** (10): 767–786.

Rubio-Rodríguez, N., Beltrán, S., Jaime, I., *et al.* (2010). Production of omega-3 polyunsaturated fatty acid concentrates: A review. *Innovative Food Science and Emerging Technologies*, **11**: 1–12.

Ruiter, A. (1995). *Fish and Fishery Products*, pp. 387, Biddles Ltd, Guildford, UK.

Saito, T., Arai, K. & Matsuyoshi, M. (1959). A new method for estimating the freshness of fish. *Bulletin of the Japanese Society of Scientific Fisheries*, **4**: 749–750.

Saldamli, I. (1998). *Gıda Kimyası*, pp. 527, Hacettepe Üniversitesi Yayınları, Ankara.

Sanchez-Alonso, I., Careche, M. & Borderias, A.J. (2007). Method for producing a functional protein concentrate from giant squid (*Dosidicus gigas*) muscle. *Food Chemistry*, **100**: 48–54.

Sanchez-Machado, D.I., Lopez-Cervantes, J., Lopez-Hernandez, J. & Paseiro-Losada, P. (2004). Fatty acids, total lipid, protein and ash contents of processed edible seaweeds. *Food Chemistry*, **85**: 439–444.

Sen, D.P. (2005). *Advances in Fish Processing Technology*, pp. 786, Allied Publishers, New Delhi.

Sharma, R. & Katz, J. (2013). Fish proteins in coronary artery disease prevention: Amino acid–fatty acid concept. *Bioactive Food as Dietary Inventions for Cardiovascular Disease*, (Eds. Watson, R.R. & Preedy, V.R.) pp. 525–249, Elsevier, New York.

Simopoulos, A.P., (2000). Symposium: role of poultry products in enriching the human diet with n-3 PUFA. *Poultry Science*, **79**: 961–970.

Simopoulos, A.P., (2002). The importance of the ratio of omega-6/omega-3 essential fatty acids. *Biomedicine and Pharmacotherapy*, **56**: 365–379.

Sivik, J. (2000). *Assessment of Protein Fingerprinting Method for Species Verification of Meats*. The Faculty of Graduate Studies of The University of Guelph, Master of Science Thesis, 171 pp.

Sorensen, N.S., Marckmann, P., Hoy, C.E., Van Duyvenvoorde, W. & Princen H.M. (1998). Effect of fish-oil-enriched margarine on plasma lipids, low-density-lipoprotein particle composition, size, and susceptibility to oxidation. *American Journal of Clinical Nutrition*, **68**: 235–241.

Spitze, A.R., Wong, D.L., Rogers, Q.R. & Fascetti, A.J. (2003). Taurine concentrations in animal feed ingredients; cooking influences taurine content. *Journal of Animal Physiology and Animal Nutrition*, **87** (7–8): 251–262.

Stancheva, M., Merdzhanova, A., Dobreva, D.A. & Makedonski, L. (2010). Fatty acid composition and fat-soluble vitamins content of sprat (*Sprattus sprattus*) and goby (*Neogobius rattan*) from Bulgarian Black Sea. *Ovidius University Annals of Chemistry*, **21** (1): 23–28.

Stelo, C.G. & Rehbein, H. (2000). TMAO-degrading enzymes. *Seafood Enzymes* (Eds Haard, N.F. & Simpson, B.K.), pp. 167–190, Marcel Dekker Inc., New York.

Suriah, A.B., Huah, T.S., Hassan, O. & Duad, N.M. (1995). Fatty acid composition of some Malaysian freshwater fish. *Journal of Food Chemistry*, **54**: 45–49.

Timm-Heinrich, M., Nielsen, N.S., Xu, X. & Jacobsen, C. (2004). Oxidative stability of structured lipids containing C18:0, C18:1, C18:2, C18:3 or CLA in *sn*2-position – as bulk lipids and in milk drinks. *Innovative Food Science and Emerging Technologies*, **5**: 249–261.

Toth, K. & Sugar, E. (1978). Fluorine content of foods and estimated daily intake from foods. *Acta Physiologica Academiae Scientiarum Hungaricae*, **51**: 361–369.

Tsape, K., Sinanoglou, V.J. & Miniadis-Meimaroglou, S. (2010). Comparative analysis of the fatty acid and sterol profiles of widely consumed Mediterranean crustacean species. *Food Chemistry*, **122**: 292–299.

Usydus, Z., Szlinder-Richert, J. & Adamczyk, M. (2009). Protein quality and amino acid profiles of fish products available in Poland. *Food Chemistry*, **112**: 139–145.

Velankar, N.K. & Govindan, T.K. (1958). A preliminary study of the distribution of non-protein nitrogen in some marine fishes and invertebrates. *Proceedings of the Indian Academy of Sciences*, **47** (4): 202–209.

Venugopal, V. (2006). Nutritional value and processing effects. In: *Seafood Processing*, pp. 425–446, CRC Press, Boca Raton.

Venugopal V. (2009). Polyunsaturated fatty acids and their therapeutic functions. In: *Marine Products for Healthcare*, pp. 51–101, CRC Press, Boca Raton.

Wood, J.D., Richardson, R.I., Nute, G.R., *et al.* (2003). Effects of fatty acids on meat quality: a review. *Meat Science*, **66**: 21–32.

Yerlikaya, P., Topuz, O.K., Buyukbenli, H.A. & Gökoğlu, N. (2013). Fatty acid profiles of different shrimp species: effects of depth of catching. *Journal of Aquatic Food Product Technology*, **22** (3): 290–297.

Zlatanos, S. & Laskaridis, K. (2007). Seasonal variation in the fatty acid composition of three Mediterranean fish sardine (*Sardina pilchardus*), anchovy (*Engraulis encrasicholus*) and picarel (*Spicara smaris*). *Food Chemistry*, **103**: 725–728.

CHAPTER 3

Quality changes and spoilage of fish

3.1 Introduction

What is quality? The term of quality has different meanings. Most often 'quality' refers to the aesthetic appearance and freshness or degree of spoilage that the fish has undergone. It may also involve safety aspects such as being free from harmful bacteria, parasites or chemicals. It is important to remember that 'quality' implies different things to different people, and is a term which must be defined in association with an individual product type. In the fishing industry the term 'quality fish' often related to expensive species or to the size of fish. Fish considered by a processor to be of inferior quality may be too small or to poor a condition for a certain process, resulting in low yields and profit. For example, it is often thought that the best quality is found in fish which are consumed within the first few hours post mortem. However, very fresh fish which are in rigor-mortis are difficult to fillet and skin, and are often unsuitable for smoking. Thus, for the processor, slightly older fish which have passed through the rigor process are more desirable (Huss 1995).

Quality usually refers to the same meaning as freshness. Freshness is the major contribution to the quality of seafood products. For all

Seafood Chilling, Refrigeration and Freezing: Science and Technology, First Edition.
Nalan Gökoğlu and Pınar Yerlikaya.
© 2015 John Wiley & Sons, Ltd. Published 2015 by John Wiley & Sons, Ltd.

kinds of seafood products, freshness is essential for the quality of final product (Alasalvar *et al.* 2010). Postharvest quality of seafood alters rapidly. Quality attributes of fish flesh – including food safety, organoleptic features, nutritional quality and aptitude to industrial transformation – influence consumption and acceptability of fish as food. Fish sensorial changes and texture properties are closely linked to freshness (Delbarre-Ladrat *et al.* 2006).

3.2 Factors affecting quality of fish

The type and number of factors influencing quality of fishery products are numerous. Postharvest biochemical and microbial changes in fish tissue depend very significantly upon the factors which effect the concentration of substrates and metabolites in the tissues of the live fish, the activity of the endogenous enzymes, the microbial contamination and the conditions after catching (Sikorski *et al.* 1990). Along with ante-mortem muscle biochemistry, post-mortem biochemical processes are directly linked to final quality attributes. The understanding of post-mortem mechanisms is a prerequisite for an accurate control of the quality of commercialized fish by the identification of objective markers or indicators (Delbarre-Ladrat *et al.* 2006).

The physical, chemical and bacteriological characteristics of fish tend to differ with species, feeding habits, season, spawning cycles, methods of catching, fishing grounds, size, age, microbiological load and geographical location (Shewan 1977; Huss 1995). Nevertheless, temperature and rigor-mortis are the main underlying factors in fish spoilage.

3.2.1 Species
The rate of spoilage is species dependent. It is well-known fact, that when chilled or frozen, fatty species such as sardine and mackerel will spoil more rapidly than a lean species like cod. The fat content of pelagic fish species can vary considerably throughout the year. Differences in composition within a species may be the cause of secondary influences on quality. When placed in refrigerated storage, lean fish in poor condition spoil much more rapidly than specimens of the same species in good condition. This may be explained by

glycogen content of the flesh. Low flesh pH also has undesirable effects on the quality of fish. The final pH of cod flesh has some influence over the rate at which deterioration occurs during frozen storage. Those species of fish caught in warm waters keep longer on ice than those caught in cooler waters. The reason for this is bacterial flora growing on the surface of fish. Another species effect is related to migratory routes. Those species migrating for long distances prior to capture will, in all probability, not be in as good physical condition as those species or specimens of the same species that follow shorter routes (Wheaton &Lawson 1985).

3.2.2 Size

Size heterogeneity develops in fish populations as the animal matures because differences exist in feeding behaviour and growth rate. Heterogeneous growth may impact composition and other muscle attributes at harvest because of developmental differences in tissue compartments at harvest (Jittinandana *et al*. 2003). Large fish keep better than small fish. One of the main mechanisms of spoilage is penetration of microorganisms from the surface to the interior of the fish. Larger fish have a smaller surface area to volume ratio so that in the same time period, less of the interior of larger fish is affected. Also, large fish such as cod, tuna and salmon are generally eviscerated aboard the fishing vessel. Flesh pH is another size effect. Small fish tend to have higher post-rigor pH than larger fish of same species (Wheaton & Lawson 1985). In general, large fish of a given species fetch the highest prices. Consumers are prepared to pay more for large samples of commodities like shrimps, scampi, crab and lobster because they are visually and gastronomically more satisfying. That is to say, large fish are not necessarily more finely flavoured and textured than small ones. Processors place a high value on large fish because the percentage yield of edible material is high, handling costs per unit weight are lower, they often keep better, and often more uniform products can be made from them. On the other hand, for some purposes the optimum size is less than the largest. Large sizes of trout, clams and oysters are not favoured for table use, because portion size is then too large or expensive. For canning, specific sizes of sprat, herring, sardine and similar species are required to ensure correct can fill (Connell 1975).

3.2.3 Distance to port

How quickly the fish are eviscerated and placed into cold storage may be related to the distance the vessel must travel from its home port to the fishing grounds. The problem of distance from fishing grounds to port is more pronounced in tropical and subtropical regions than it is in the colder climates. The heat from the sun quickly overheats the fish and accelerates post-mortem changes (Wheaton & Lawson 1985).

3.2.4 Diet of fish

Heavily feeding fish tend to be more susceptible to autolytic tissue degradation than the petite feeders. The type of feed/food on which fish is feeding on may similarly have an effect on their spoilage rate during storage. Non-feeding fish have been found to have low levels of bacteria in the intestines as compared to the heavily feeding fish (Huss 1988).

The chemical composition of fish varies greatly depending on feed, environment and season. It is reported that the lipid content of farmed common carp can vary widely depending on the feed used. The quantity of n-3 fatty acids varies largely with the origin of diets and its composition (Ljubojević *et al.* 2012).

Fattiness in cultured fish generally is undesirable because consumers consider eating fish as part of a low-fat diet. Fat in fish may not only reduce consumer acceptance, but also may reduce storage time and decreased processed yield of the fish. There are several factors that affect the amount of fat in fish such as feeding (Lovell 1998). Diet of fish has significant effect on flavour of its flesh. Some freshwater fish suffer from a muddy odour and flavour, which may reduce consumer acceptance (Wheaton & Lawson 1985).

3.2.5 Fishing grounds and methods

The fishing grounds and geographical locations exclusively tend to determine fish spoilage in so far as temperature, food types, the level of pollution; microenvironments are concerned (Shewan 1977) The location of the fishing grounds plays an indirect role in determining the quality of fishery product. Within a species, flavour can vary from one ground to the next, depending upon the nature of food and physiological condition of species of interest. Winds, tides water conditions and migratory patterns also have some influence on the condition and quality of the before harvest. The time of year when spawning and subsequent poor condition develops in fish varies with the fishing grounds (Wheaton & Lawson 1985). Geographical location, and or the type of waters, has a remarkable influence on the type of micro fauna that will grow on the fish. The microfauna found in fish from warm waters (tropical waters) differs from those from cold waters. The predominant bacteria on the surface of fish in warm waters consist of the Gram-positive bacteria; *Bacillus, Micrococcus* and *Carnobacterium*, which constitute 50–60% of the total microflora. Those found in cold waters comprise Gram-negative bacteria; *Psychrobacter, Moraxella, Pseudomonas* ssp, *Actinobacter, Shewanella, Flavobacterium, Cytophaga* and *Vibrio* (Gram *et al.* 1989). The common spoilage organisms associated with seafood stored in ice and air, whether of temperate, sub-tropical or tropical origin, are the Gram-negative psychrotrophic bacteria such as *Pseudomonas* and *Shewanella putrefaciens*. Other pertinent spoilage bacteria include: *P. phosphoreum*, commonly isolated in packed fish and other meats; Vibrionaceae, associated with fish at elevated temperatures anaerobic spoilers such as *Lactobacillus* spp. and *Leuconostoc* spp. (Huss 1995).

The method of fishing, as well, has implications for the glycogen level at the time of fish death; more exhausted fish, for instance those caught by trawl method exhibit lower glycogen reserves in the muscle cells thus, higher ultimate pH at the resolution of rigor mortis and therefore, more prone to higher rate of spoilage as high pH favours microbial proliferation (Huss 1995).

3.2.6 Sex

Sex plays a large role in quality soon after spawning. The females of certain species may be in such poor physical condition soon after spawning that they are of very poor quality. However, in some species such as salmon, both sexes may be in poor condition after spawning. Just prior to and during spawning, food reserves in the flesh are transferred for the development of the gonads. During spawning, most fish do not feed. In this case the flesh becomes severely depleted of fat, protein and carbohydrates and the fish are in poor condition (Wheaton & Lawson 1985). Spawning fish tend to use most of their glycogen, and the effect of depletion is reflected in their susceptibility to rapid deterioration during storage due to faster onset and resolution of rigor mortis (Connell 1990; Huss 1995).

3.3 Post-mortem changes in fish muscle

Fish and fish products are fast0deteriorating or perishable materials. The easy deterioration in fish quality is because of the post-mortem biological changes that take place in the body of dead fish. The biochemical post-mortem changes, involving the main chemical constituents of the tissues, bring about various structural alterations in the tissues, including rigor-mortis, and different degrees of disintegration of the muscle ultrastructure. The degradation of the different constituents of the skin and muscles leads to gradual development of staleness and spoilage of fish and shellfish (Sikorski *et al.* 1990). Changes in fish quality often come in the form of unpleasant odour and microbial spoilage. Post-mortem changes that take place in fish tissue occur in the following phases: slime secretion on the surface of fish, rigor mortis, autolysis as enzymatic decomposition of tissues, and microbiological spoilage (Wakjira 2011). Post-mortem changes and rigor-mortis formation are summarized in Figure 3.1.

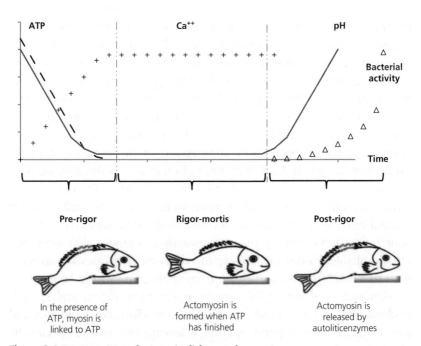

Figure 3.1 Post-mortem changes in fish muscle.

3.3.1 Rigor mortis

Rigor or rigor mortis means the stiffening of the muscles of an animal shortly after death. Immediately after death the muscles of an animal are soft and limp, and can easily be flexed; at this time the flesh is said to be in the pre-rigor condition, and it is possible to make the muscles contract by stimulation, for example by means of an electric shock. Eventually, the muscles begin to stiffen and harden, and the animal is then said to be in rigor. The muscles will no longer contract when stimulated, and they never regain this property. After some hours or days the muscles gradually begin to soften and become limp again. The animal has now passed through rigor, and the muscle is in the post-rigor condition. Sometimes rigor is said to be resolved; this is simply another way of saying that the muscle has passed through rigor to the post-rigor stage (Stroud 1969).

The stiffness of a fish in rigor is a sure sign of freshness. However, it may have an adverse effect on the suitability of fish for mechanical or even hand filleting. After several hours, the rigid fish gradually softens and becomes pliable again, although the extension of muscles under stress is not reversible and the muscle has no ability to respond to electrical stimuli. The biochemical state of the post-rigor muscles is different from that of the pre-rigor flesh. The post-rigor fish does not display the signs of prime freshness (Sikorski *et al.* 1990) While the fish is alive, cycles of chemical changes take place continuously in the muscle; these provide energy for the muscle while the fish is swimming, and also produce substances necessary for growth and replacement of worn-out tissue. The compounds that bring about, and control, these changes are known as enzymes. The enzymes in the flesh go on working even after the fish is dead, and some of them act on those substances that normally keep the muscle pliable and lifelike. During life the muscle would contract and become rigid if its two main protein components were allowed to interact and bond together, but the bonding is prevented by the presence of substances that keep the muscle pliable, rather in the way in which oil lubricates the moving parts of a machine and prevents it from seizing up (Stroud 1969).

Generally, stiffness in fish begins in the head region and spreads gradually towards the caudal muscles. The time a fish takes to go into, and pass through; rigor depends on the following factors: the species, its physical condition, the degree of exhaustion before death, its size, the amount of handling during rigor and the temperature at which it is kept. Some species take longer than others to go into rigor, because of differences in their chemical composition. The poorer the physical condition of a fish – that is the less well nourished it is before capture – the shorter will be the time it takes to go into rigor; this is because there is very little reserve of energy in the muscle to keep it pliable. Fish that are spent after spawning are an example. In the same way, fish that have struggled in the net for a long time before they are hauled aboard and gutted will have much less reserve of energy than those that entered the net just before hauling, and thus will go into rigor more quickly. Small fish usually go into rigor faster than large fish of the same species. Manipulation of pre-rigor fish does not appear to affect the time of onset of rigor, but manipulation, or flexing, of the fish while in rigor can shorten the time they remain stiff. Temperature is perhaps the most important factor governing the time a fish takes to go into, and pass through, rigor because the temperature at which the fish is kept can be controlled. The warmer the fish, the sooner it will go into rigor and pass through rigor (Stroud 1969).

The quality of fish, regardless of the species characteristics, is affected both by the freshness and appearance of the fillet. As long as the fish is in the pre-rigor or rigor state, its freshness is impeccable. The biochemical and biophysical processes involved in rigor may influence the appearance of the fillet and the amount of liquid lost due to freezing/ thawing and cooking. Under the contraction associated with rigor mortis, some myosepta connecting the myotomes of the muscles, break, yielding to the tension, and the flakes come apart. The gaping is severe if the tension is strong and the connective tissue is weakened, especially when rigor sets in at high ambient temperature. Rough handling of fish in rigor increases the incidence of gaping. Fillets cut from well-nourished fish just after catch shrink at high ambient temperature due to rigor mortis by as much as 30 to 40% of initial length. Fish entering rigor at high ambient temperatures loses much drip on thawing and may be tough and stringy after cooking (Sikorski et al. 1990).

3.3.2 Chemical changes

In muscle tissue of fish, autolytic and proteolytic changes catalysed by microbial enzymes take place during storage. Proteins are being gradually cleaved into peptides, amino acids, ammonia and some other low molecular N-substances. Toxic biogenic amines (histamine, tyramine)

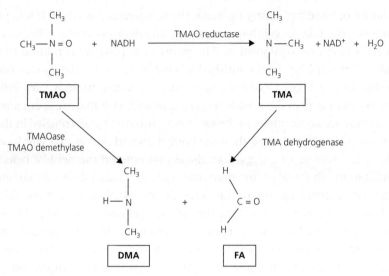

Figure 3.2 Degradation of trimethylamineoxide (TMAO).

may be produced by activities of some microorganisms. The shelf life of fat fish is limited by lipid decomposition (Ježek & Buchtová 2007).

The deteriorative changes occurring in fish result in the gradual accumulation of certain compounds in the flesh. Quantification of these compounds can provide a measure of the progress of deterioration (Connell 1995). Among the chemical indices of spoilage assessed are trimethylamine (TMA), total volatile bases (TVB) and hypoxanthine contents of the flesh. TMA is the best known compound produced during fish spoilage and it is mainly derived from bacterial breakdown of trimethylamine oxide (TMAO) (Figure 3.2). TVB content is an alternative to measuring TMA, and includes ammonia, dimethylamine (DMA) and TMA. Amino acids are important bacterial substrates for the formation of sulphides and ammonia. Total volatile basic nitrogen (TVB-N) and trimethylamine (TMA-N) content are the most chemical parameters used for determination fish quality. The levels of these compounds increased with the onset of spoilage. This chemical compound is the primary cause for the fishy odours, which increased as spoilage proceeds and show good correlation with sensory analysis (Farag 2012).

The chemical spoilage associated with fish during storage is mainly due to fish lipid degradation (auto-oxidation). In general, fish have high degree of unsaturated lipids than other food commodities (Huss 1995). The oils and unsaturated fatty compounds contained in fish flesh and other tissue can undergo changes while the fish is being stored that produce rancid odours, off-flavours and colour changes. Fish normally have much higher degree of lipid un-saturation than most other foods and, therefore, are particularly susceptible to oxidative rancidity. The species is the most important factor determining the rate of rancidity in fish. Fish that have a high fat content have relatively short frozen storage life because of the susceptibility of fish to oxidative rancidity (Wheaton & Lawson 1985). Fish lipids are subjected to two main changes: lipolysis and auto-oxidation. The main reactants in these processes involves atmospheric oxygen and fish unsaturated lipids, leading to the formation of hydroperoxides, associated with tasteless flavour and accompanied by brown–yellow discoloration of the fish tissue (Huss 1994). Fat oxidation usually occurs after autolysis and bacterial spoilage. The lipid concentration in fish can contribute to the spoilage process in fish. The fats in fish are mainly unsaturated fatty acids that are easily oxidised by oxygen from the atmosphere. High temperature or exposure to light can increase the oxidation rate. For fatty fish preserved in ice, spoilage due to rancidity is mainly caused by oxidation. This produces a bad and unpleasant odour as well as a rancid taste (Hobbs 1982).

Lipid oxidation is one of the major problems in the fish industry, due to flavour deterioration and nutritional value loss (Taheri & Motallebi 2012). Fatty fishes are very sensitive materials due to their high dark muscle and fat contents. So they rapidly lose their freshness and the functional properties of proteins. Lipids in fatty fish are rich in long-chain polyunsaturated fatty acids of excellent nutritional value. However, they are very prone to oxidation. Myoglobin and oxidizing enzymes present in dark fish muscle are efficient pro-oxidants. Lipid oxidation induces formation of an array of products, which decrease the sensory quality of fish and fish products directly or indirectly (Eymard et al. 2005). Oxidation is influenced by handling and processing of fish. Oxidation process can lower nutritional quality and modify texture and colour. Reactions are initiated and accelerated by heat, light, (especially UV-light) and several inorganic and organic substances (such as copper and iron ions) and several antioxidants with the opposite effect (such as α-tocopherol, ascorbic acid, citric acid, carotenoids) (Huss 1994).

Lipid hydrolysis is an enzymatic reaction catalysed by lipase, and the free fatty acid content rises during storage. Glycerides, glycolipids and phospholipids are hydrolysed by lipases to free fatty acids, which produce low-molecular-weight compounds (such as aldehydes and ketones) in further oxidation. These compounds are responsible for off-flavour and off-odour and taste of fish. Lipid hydrolysis favours lipid oxidation because the fatty acids formed can be substrates of the oxidation reaction (Gökoğlu et al. 2012).

3.3.3 Microbiological changes

Bacteria are capable of causing spoilage because of two important characteristics. First, they are psychotropic and thus multiply at refrigeration temperatures. Second, they attach various substances in the fish tissue to produce compounds associated with off-flavours and off-odours. When the fish is alive the bacteria are found on the gill and skin and in the intestines, but cannot attack the fish muscle. However, when the fish dies the bacteria can penetrate into the flesh muscle of the fish. When fish is preserved by icing the rate of bacterial penetration into the flesh muscle is much slower. Fish spoilage occurs when the enzyme of bacteria diffuses into the flesh muscle and the nutrition substances from the flesh muscle diffuse to the outside. Spoilage will happen more rapidly for fish species with a thin skin layer (Gram & Huss 1996). Microorganisms are found on all the outer surfaces (skin and gills) and in the intestines of live and newly caught fish. The bacterial flora on newly caught fish depends on the environment in which it is caught rather than on the fish species. Fish caught in very cold, clean waters carry lower numbers whereas fish caught in warm waters have slightly higher counts. Very high numbers are found on fish from polluted warm waters. Many different bacterial species can be found on the fish surfaces. The bacteria on temperate water fish are all classified according to their growth temperature range as either psychrotrophs or psychrophiles. Psychrotrophs (cold-tolerant) are bacteria capable of growth at 0°C but with an optimum around 25°C. Psychrophiles (cold-loving) are bacteria with maximum growth temperature around 20°C and an optimum temperature at 15°C. In warmer waters, higher numbers of mesophiles can be isolated. The microflora on temperate water fish is dominated by psychrotrophic Gram-negative rod-shaped bacteria belonging to the genera Pseudomonas, Moraxella, Acinetobacter, Shewanella and Flavobacterium. Members of the

Vibrionaceae (*Vibrio* and *Photobacterium*) and Aeromonadaceae (*Aeromonas* spp.) are also common aquatic bacteria and typical of fish. Gram-positive organisms as *Bacillus*, *Micrococcus*, *Clostridium*, *Lactobacillus* and coryneforms can also be found in varying proportions, but in general, Gram-negative bacteria dominate the microflora (Gram & Huss 1996).

There are many bacteria species present in spoiling fish but there are only certain types that are considered to cause spoilage. The bacteria use their enzyme to change fish odour and flavour to sour, gassy, fruity and finally ammonia and faecal odour appear (Quang 2005). The spoilage pattern of fish may be influenced by various parameters, such as the microbiota acquired during handling and processing, the conditions to which the fish has been subjected during storage and processing, and on the actual chemical composition of the fish. Non-bacterial spoilage is more important in fatty fish where oxidation of the lipids produces off-odours and off-flavours (Lauzon *et al.* 2010).

Since fish will, to a large degree, reflect the microfloras of the environment from they came, fish from inland and coastal waters are subjected to varying degrees of pollution. They may be contaminated with organisms such as *Salmonella*, *Shigella* and *Escherichia coli*. Microbial growth begins with the death of the fish, at which time the natural defence mechanisms of fish are destroyed. The rate of growth depends upon the number of the types of microorganisms present on the fish and the temperature at which the fish is being held. Bacteria are able to produce quite rapidly under the right conditions. To produce, a bacterial cell splits in half, and each half grows into a new complete cell through a process called 'fission'. The time it takes for bacteria for reproduce is dependent upon temperature. At room temperature, most spoilage bacteria will divide approximately every 30 minutes. A remarkable number of individuals can be produced from a single bacterium within a few hours. Reproduction and growth rates of bacteria generally increase at temperatures above 4°C. When considering fish spoilage, it is important to consider also the psychrophilic nature of microorganisms. Psychrophilic organisms are capable of reproducing to large numbers at 0°C and higher. In regard to fish spoilage psychrophilic, to other important groups of bacteria are mesophiles and thermophiles (Wheaton & Lawson 1985).

The knowledge of bacterial spoilage of freshwater fish is poor compared with that for marine fish; however, there are similarities in the spoilage patterns. As in marine fish, bacteria and their metabolic productsare responsible for spoilage in addition to enzymes in the fish muscle and intestines. Freshwater fish typically carry freshwater bacteria. These include most genera found in sea water, plus species of *Aeromonas, Lactobacillus, Brevibacterium, Alcaligenes* and S*treptococcus*. The intestines of both marine and freshwater fish contain bacteria of the genera *Achromobacter, Pseudomonas, Flavobacterium, Vibrio, Bacillus, Clostridium* and *Escherichia* (Wheaton & Lawson 1985).

Shellfish are generally subject to types of microbial spoilage similar to those occurring in fish. Shrimps, crabs, lobsters, and similar aquatic creatures, have a bacterial slime on their surfaces similar to fish. Oysters, clams and other shellfish that filter large amounts of water through their bodies pick up soil and water microorganisms, including pathogens. *Achromobacter* and *Flavobacterium* predominate in these animals. Oysters remain in good conditions as long as they are kept under refrigeration in the shell, but once they are removed from the shell they deteriorate rapidly (Wheaton & Lawson 1985).

3.3.4 Enzymatic changes

The initial quality loss in fish is basically due to autolytic changes and is not related to microbiological activity. The autolytic spoilage changes precede the other changes responsible for the loss of fish quality during storage. Autolysis is principally enzymatic reactions, which take place in fish tissues. Enzymes and other related chemical reactions do not immediately cease their activities in the fish muscle after fish death. Their continuation initiates other processes like rigor-mortis; which is the basis for autolytic spoilage in fish (Huss 1995).

Enzymes are protein-like substances present in the flesh and stomach of fish and shellfish which initiate or speed up chemical reactions. When the fish is alive, enzymes are usually kept in balance with the help of digestive or blood systems. They remain active after the death of the fish and are particularly involved in flavour changes that take place during the first few days of storage before bacterial spoilage becomes significant. In a short time enzymatic activity can also alter the texture and appearance of the flesh (Wheaton & Lawson 1985). Autolysis is the breakdown or decomposition of larger molecules such as proteins, lipids and carbohydrates under the influence of enzymes up on the fish death. The quality of fish as a raw material for consumption or for processing depends largely on proteolysis, which is the autolysis of proteins (Wakjira 2011). When captured or harvested, fish and shellfish usually contain food in their gut, and powerful enzymes are present. Upon the death of the animal, the enzymes penetrate the gut wall and surrounding flesh, weakening and softening them. The gut and flesh may then be invaded by spoilage bacteria (Wheaton & Lawson 1985).

Enzymes play a role in the development of rigor mortis, which is the progressive stiffening of muscles several hours after death. The stiffening effect is as a result of coagulation of muscle protein. The duration of intensity of rigor depends upon the species, temperature and condition of fish. It usually passes before bacteria invade the muscle, leaving the flesh soft and limp. Following rigor the self-digestion process commences, as a result of enzymatic activity. The gut enzymes are particularly active at this time. A phenomenon known as 'burst belly' can occur in just a few hours in some fish, such as sardine and herring, and is caused by weakening of the belly wall due to self-digestion (Wheaton & Lawson 1985).

The initial quality loss is mainly explained by autolytic changes, such as degradation of nucleotides (ATP-related compounds) by autolytic enzymes. The loss of the intermediate nucleotide, inosine monophosphate (IMP), is responsible for the loss of fresh fish flavour. These autolytic changes make catabolites available for bacterial growth (Huss 1995). Most important in the autolytic spoilage changes is the degradation of fish nucleotides. Adenosine triphosphate (ATP) degrades to adenosine diphosphate (ADP), adenosine monophosphate (AMP), IMP, Inosine (Ino) and hypoxanthine (Hx), associated with bitter fish flavours. Generally, the biochemical changes due to enzymatic activity related to freshness deterioration in fish are change in the flavour and colour. The ultimate degradation of ATP intermediate nucleotides (IMP and inosine) to Hx is attributed to bacterial activity (Haard 2002). However, the presence of ATP intermediate nucleotides, monophosphate (IMP) and inosine (Ino) is associated with the desirable sweet flavour in fresh fish (Huss 1995). The ATP of fish muscle breaks down, either during the death struggle, or subsequently. This breakdown results in liberation of IMP, the contributor of the pleasant flavour of fresh fish. The degradation of IMP to hypoxanthine is a factor in the progressive loss of desirable flavour and in the development of bitter off-flavour (Mandal & Mukherjee 1974). The first autolytic processes in the fish muscle tissue involve the carbohydrates and the nucleotides. For a short period, the muscle cells continue the normal physiological processes but soon the production of ATP stops. ATP functions as an ubiquitous energy donor in numerous metabolic processes. In the live organism, ATP is formed by the reaction between ADP and creatine phosphate, the latter being a reservoir of energy-rich phosphate in the muscle cells. When the reservoir is depleted, the ATP is regenerated from ADP by re-phosphorylation during glycolysis. After death, when the regeneration ceases, the ATP is rapidly degraded. At low ATP levels rigor-mortis develops (Huss 1988). It is well known that enzymes from the digestive tract play an important role in the autolysis of whole, uneviscerated fish. During periods of heavy feeding the belly of certain fish is very susceptible to tissue degradation and may burst within a few hours of catch (Huss 1988). Production of lactic acid tends to lower pH towards 6.4, depending on species, causing liberation and activation of tissue proteases. Proteolysis may result in an increase in free amino acids but the rate seems to vary markedly with species. The most noticeable physical change resulting from autolysis in the early stages of deterioration is bursting of the belly walls of ungutted fish that have been eating heavily. The concentration and activity digestive enzymes are highly in the guts and soon being to digest the gut walls and surrounding tissues (Whittle *et al.* 1990).

Glycolysis is the anaerobic formation in muscle of lactate from glycogen by glycolysis, which continues for a time to provide energy to maintain muscle functions. Glycolytic changes are rapid and the rate is considerable even at temperatures just below 0°C. In general, fish muscles contain a relatively low amount of glycogen compared with mammalian muscle and the final post-mortem pH is consequently higher; this makes fish meat more susceptible to microbial attack. However, there are great variations in the glycogen content of different species; for example, tuna have levels comparable with those found in mammals, and also within the same species. As a rule, well-rested fish contain more glycogen than exhausted fish, well fed more than starved fish and large more than small fish. Within the fish, glycogen is more concentrated in the dark muscle than in the white muscle (Wakjira 2011). Figure 3.3 shows glycogen breakdown after death.

Among postharvest changes, degradation of fish muscle caused by endogenous proteases is a primary cause of quality losses during cold storage of handling (Haard 1994). Proteases are able to hydrolyse the

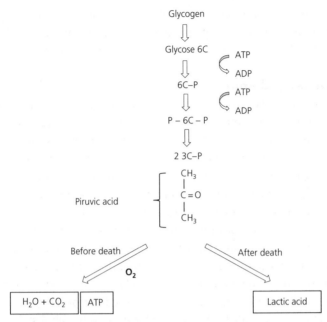

Figure 3.3 Glycogen breakdown after death of fish.

muscle proteins. Microbial proteases may also be a potential cause of proteolytic degradation. Proteolytic enzymes are found in all tissue, although both the distribution of different enzymes and their activities show considerable variation. The highest activities are found in fractions such as viscera and liver, but there are significant proteolytic activities in muscle tissue as well, where the enzymes play an important role in protein turnover. After death, the biological regulation of the enzymes is lost, and the enzymes hydrolyze muscle proteins and resolve the rigor-mortis contraction (Hultmann 2003).

Post-mortem lipid degradation proceeds mainly due to enzymatic hydrolysis. Free fatty acids (FFA) accumulate in muscle lipids from enzymatic hydrolysis of lipids, and they degrade the quality of fish muscle. Lipid hydrolysis is an enzymatic reaction catalysed by lipase and free fatty acid content arises during storage.

Some crustaceans such as shrimp, lobster suffer from an enzymatic change which results in a condition known as 'black spot' or 'melanosis'. This condition is caused by build-up of black pigment beneath the sell, occurring first in the membranes that connect the overlapping ends of tail segments. Black spot does not affect the edibility of such shrimp but it is sign of poor handling aboard the realer and the shrimp are considered of poorer quality (Wheaton & Lawson 1985).

3.3.5 Sensory changes

During fish spoilage, there is a breakdown of various components and the formation of new compounds. These new compounds are responsible for the changes in odour, flavour and texture of the fish meat (Ghaly *et al.* 2010). The quality changes can easily be noticed and consist of changes in colour, odour or smell, taste, appearance and texture, and are therefore called sensory changes. After harvest chemical and biological changes occurs in fish due to enzymatic breakdown. Autolytic enzymes reduce the textural quality of fish. A number of proteolytic enzymes are found in muscle and viscera of the fish after catch. These enzymes contribute to post mortem degradation in fish muscle and fish products during storage and processing. There is a sensorial or product associated alteration that can be contributed by proteolytic enzymes (Engvang & Nielsen 2001). Microbial growth and metabolism is a major cause of fish spoilage, which produces amines, biogenic amines, organic acids, sulphides, alcohols, aldehydes and ketones, with unpleasant and unacceptable off-flavours (Ghaly *et al*, 2010). The first sensory changes of fish during storage are concerned with appearance and texture. Sensory properties such as appearance, odour, taste and texture of fish are the most important factors for consumer choice.

Many factors such as species, age, size and nutritional state of the fish affect the texture of fish. Post-mortem factors influencing texture include glycolysis, pH and rigor mortis. The accompanying contraction of the muscle often leads to the separation of muscle segments (gaping). External factors include the temperature profile during storage, temperature of cooking, and the presence of NaCl. As the fish grows, the diameter and length of the muscle fibres increases, and this makes the muscle coarser. One of the most dramatic changes in fish muscle post mortem occurs when it passes through rigor mortis. Rigor mortis has a major effect on texture, especially in fish frozen-at-sea. Sea-frozen fish is often filleted and frozen pre-rigor and this can lead to texture damage. A slight or more dramatic decrease in pH can be observed during rigor mortis in most species, because lactic acid is formed from glycogen. Variations in pH depend on many factors (Hyldig & Nielsen 2007).

References

Alasalvar, C., Grigor, J.M. & Ali, Z. (2010). Practical evaluation of fish quality by objective, subjective and statistical testing. In: *Handbook of Seafood Quality, Safety and Health Applications* (Eds Alasalvar, C., Shahidi, F., Miyashita, K. & Wanasundara, U.), pp. 13–28, Wiley-Blackwell, Oxford.

Connell, J.J. (1975). Intrinsic quality. In: *Control of Fish Quality* 4th edn, (Ed. Connell, J.J), pp. 4–30, Fishing News Books Ltd, Oxford.

Connell, J. J. (1990). Methods of assessing and selecting for quality. In: *Control of Fish Quality*, 3rd edn, (Ed. Connell, J.J), pp. 122–150, Fishing News Books, Oxford.

Connell, J.J. (1995). Quality deterioration and extrinsic quality defects in raw material. In: *Control of Fish Quality* 4th edn, (Ed. Connell, J.J), pp. 31–35, Fishing News Books Ltd, Oxford.

Delbarre-Ladrat, C., Chéret, R., Taylor, R. & Verrez-Bagnis, V. (2006). Trends in postmortem aging in fish: understanding of proteolysis and disorganization of the myofibrillar structure. *Critical Reviews in Food Science and Nutrition*, **46** (5): 409–421.

Engvang, K. & Nielsen, H.H. (2001). Proteolysis in fresh and cold-smoked salmon during cold storage: Effects of storage time and smoking process. *Journal of Food Biochemistry*, **25**: 379–395.

Eymard, S., Carcouët, E., Rochet, M.J., Dumay, J. Chopin, C. & Genot, C. (2005). Development of lipid oxidation during manufacturing of horse mackerel surimi. *Journal of the Science of Food and Agriculture*, **85** (10): 1750–1756.

Farag, H.E.M. (2012). Sensory and chemical changes associated with microbial flora of *Oreochromis niloticus* stored in ice. *International Food Research Journal*, **19**(2): 447–453.

Ghaly, A.E., Dave, D., Budge, S. & Brooks, M.S. (2010). Fish spoilage mechanisms and preservation techniques: review. *American Journal of Applied Sciences*, **7** (7): 859–877.

Gökoğlu, N., Yerlikaya, P., Topuz, O.K. & Buyukbenli, H.A. (2012). Effects of plant extracts on lipid oxidation in fish croquette during frozen storage. *Food Science and Biotechnology*, **21** (6): 1641–1645.

Gram, L., Oundo, J. & Bon, J. (1989). Storage life of Nile perch (*Lates niloticus*) dependent on storage temperature and initial bacterial load. *Tropical Science*, **29**: 21–236.

Gram L. & Huss H.H. (1996). Microbiological spoilage of fish and fish products.*International Journal of Food Microbiology*, **33**: 121–137.

Haard, N.F. (1994). Protein hydrolysis in seafoods. In: *Seafoods: Chemistry, Processing Technology and Quality* (Ed. Shahidi, F. & Botta, J.R.), pp. 10–33. Chapman &Hall, New York.

Haard, N. (2002). The role of enzymes in determining seafood color and texture. In: *Safety and Quality Issues in Food Processing* (Ed. A.H. Bremner), pp. 221–253, Woodhead Publishing Limited, Boca Raton, Boston, New York, Washington, DC.

Hobbs G. 1982. Changes in fish after catching. *Fish Handling and Processing.* Torry Research Station, pp. 20–27.

Hultmann, L. (2003). Endogenous proteolytic enzymes - Studies of their impact on fish muscle proteins and texture. Doctoral thesis 2003:110. Faculty of Natural Sciences and Technology Department of Biotechnology, Norwegian University Science and Technology.

Huss, H.H. (1988). *Fresh fish quality and quality changes.* FAO/DANIDA. A training manual, FAO Fisheries Series No. 29, Food and Agriculture Organization of the United Nations, Rome, 1988.

Huss, H.H. (1994). Assurance of seafood quality. FAO Fisheries Technical Paper 334. Food and Agriculture Organization of the United Nations, Rome, 1994.

Huss, H.H. (1995). Quality and quality changes in fresh fish. FAO Fisheries Technical Paper 348. Food and Agriculture Organization of the United Nations.Rome, 1995.

Hyldig, G. & Nielsen, D. (2007). A review of sensory and instrumental methods used to evaluate the texture of fish muscle. *Journal of Texture Studies,* **32**: 219–242.

Ježek, F. & Buchtová, H. (2007).Physical and chemical changes in fresh chilled muscle tissue of common carp (*Cyprinus carpio* L.) packed in a modified atmosphere. *Acta Veterinaria Brno,* **76**: 83–92.

Jittinandana, S., Kenney, P.B., Slider, S.D., Mazik, P., Bebak-Williams, J. & Hankins, J.A. (2003). Effect of fish attributes and handling stress on quality of smoked arctic char fillets. *Journal of Food Science,* **68**(1): 57–63.

Lauzon, H.L., Margeirsson, B., Sveinsdóttir, K., Guðjónsdóttir, M., Karlsdóttir, M.G. & Martinsdóttir, E. (2010). Overview on fish quality research - impact of fish handling, processing, storage and logistic on fish quality deterioration. Technical Report 39–10, Matis, Reykjavik, Iceland.

Ljubojević D., Ćirković, M., Novakov, N., *et al.* (2012). The impact of diet on meat quality of common carp. *Archiva Zootechnica,* **15**(3): 69–74.

Lovell, T. (1998) Dietary requirements. In: *Nutrition and Feeding of Fish* (Ed. Lovell, T.), pp. 164–170, Kluwer Academic Publishers, Boston.

Mandal, S.K. & Mukherjee, S.K. (1974). Chemical changes of fish muscle during preservation with ammonia. *Journal of Agricultural and Food Chemistry,* **22**(5): 832–835.

Quang, N.Y. (2005). Guidelines for handling and preservation of fresh fish for further processing in Vietnam. Final project. The United Nations University, Fisheries Training programme. Reykjavik, Iceland.

Shewan, J.M. (1977). The bacteriology of fresh and spoiling fish and biochemical changes induced by bacterial action. Proceedings of the Conference on Handling Processing and Marketing of Tropical Fish, Tropical products, pp. 51–66, Tropical Products Institute, London.

Sikorski, E.Z., Kolakowska, A., Burt, J.R. (1990). Postharvest biochemical and microbial changes. In: *Seafood: Resources, Nutritional Composition, and Preservation* (Ed. Sikorski, Z.E.), pp. 56–72, CRC Press Inc., Boca Raton, FL.

Stroud, G.D. (1969). Rigor in fish. The effect on quality. Torry Advisory Note 36. Torry Research Station, HMSO Press, Edinburgh.

Wakjira, M. (2011). A Distance Course of Module on Fisheries and Aquaculture (Biol 421). (Ed T. Habtamu), pp. 154, Jimma University, Ethiopia.

Taheri, S. & Motallebi, A.A. (2012). Influence of vacuum packaging and long term storage on some quality parameters of cobia (*Rachycentron canadum*) fillets during frozen storage. *American-Eurasian Journal of Agricultural and Environmental Sciences*, **12**(4): 541–547.

Wheaton, F.W. & Lawson, T.B. (1985). Quality changes in aquatic food products. In: *Processing Aquatic Food Products* (Eds Wheaton, F.W. & Lawson, T.B.), pp. 225–270. John Wiley & Sons Inc., New York.

Whittle, K.J., Hardy, R. & Hobbs, G. (1990). Chilled fish and fishery products. In: *Chilled Foods* (Ed. Gormley, T.R.), pp. 87–116, Elsevier Science Publishers, London.

CHAPTER 4

Chilling

4.1 Fundamentals of chilling

Chilling can be defined as reducing the temperature of a product to 0°C that is to the melting point of ice. Chilling at the temperature just above freezing point of the fish does not stop spoilage but retards it. Chilling is an effective method of preserving food over a short period, because it results in the retention of the main quality attributes of food. The purpose of chilling is to prolong the shelf life of fish, which it does by slowing the action of enzymes and bacteria, and the chemical and physical processes that can affect quality. Fresh fish is an extremely perishable food and deteriorates very rapidly at normal temperatures. Reducing the temperature at which the fish is kept lowers the rate of deterioration. During chilling the temperature is reduced to that of melting ice, 0°C (Shawyer & Pizalli 2003).

The temperature of the air over the product significantly affects the chilling rate. In the most suitable chilling process air temperature is first reduced to near 10°C, then rapidly to near 0°C. It should be held at this point until the majority of the heat has been extracted from the product.

Seafood Chilling, Refrigeration and Freezing: Science and Technology, First Edition.
Nalan Gökoğlu and Pınar Yerlikaya.
© 2015 John Wiley & Sons, Ltd. Published 2015 by John Wiley & Sons, Ltd.

Chilling storage is widely used because it generally results in effective short-term preservation by retarding growth of microorganisms, post-harvest metabolic activities, deteriorative chemical reactions and moisture loss. Control of temperature in the storage facility is important because it has a profound effect on growth on microorganisms, metabolic activities, chemical reactions and moisture loss. Growth of pathogenic microorganisms occurs rapidly in the range 10–37°C but only slowly in the range 3.3–10°C. Below 3.3°C pathogenic microorganisms can no longer grow. Growth of mesophilic and thermophilic microorganisms is greatly retarded at chilling temperature. Psychrotrophic microorganisms grow well in the range 0–15°C. Chilling temperatures therefore substantially retard spoilage caused by microorganisms (Karel & Lund 2003).

The choice of chilling unit depends primarily on the type of food, whether liquid, solid or semisolid. Chilling can take place by conduction, convection, radiation or evaporative cooling. Use of radiation chilling in food processing is limited. Conduction may be used when the product geometry is suitable for contact with solid chiller elements. Most food chilling relies on connective heat transfer to cool the product. The rate of cooling is determined by the rate of heat transfer through the product itself and the rate heat transfer from the product surface to coolant medium. Parameters that influence the rate of heat transfer include: surface area for heat exchange; temperature difference between product and coolant medium; thermal properties of the food; and degree of connective heat transfer due to coolant flow. The amount of heat to be removed depends on the initial temperature of the product and any heat-generation sources within the food (Heldman & Hartel 1997). Heat always moves from warmer to colder areas; it seeks a balance. If the interior of an insulated fish hold is colder than the outside air, the fish hold draws heat from the outside. The greater the temperature difference, the faster the heat flows to the colder area.

Chilling rate depends on external and internal factors. External factors are: the nature of the chilling medium (air, gases such as nitrogen, iced water); the temperature of the chilling medium; the geometric shape of the food package or container; the temperature differential between food and the chilling medium. Internal factors are heat conductivity, heat capacity, density, initial temperature, bulk and total volume, and moisture content of food (Light & Walker 1990).

4.2 Chilling of fish

Fish and other seafood spoil very quickly if they are not handled and treated properly. Spoilage of fish occurs for two main reasons: microbial spoilage and autolytic spoilage. Various techniques to preserve the freshness of fish are used. The four key factors to monitor the freshness of fish are temperature, time, care and hygiene. It is possible to preserve fresh raw fish so that it remains fresh, just like when it was caught, for a long period of time. Temperature is a critical factor to preserve fish, because the agents that cause spoilage (enzymes and bacteria) like to live in these warm conditions. It is well known that high temperatures increase the rate of fish spoilage and low temperatures slow it down. By providing low temperatures, conditions are induced that enzymes and bacteria do not like; they therefore react more slowly, and the spoiling procedure will also slow down. The faster a lower temperature is attained during fish chilling, the more effectively the spoilage activity is inhibited. The speed with which bacteria grow depends on temperature. Indeed, temperature is the most important factor controlling the speed at which fish go badly. The higher the temperature, the faster the bacteria multiply, using the flesh of the dead fish as food. When the temperature is sufficiently low, bacterial action can be stopped (Graham *et al.* 1992). Lowering the ambient temperature slows down the spoiling procedure and thus preserves the freshness of the fish for longer. Every chemical reaction needs a critical temperature. Bacterial, enzymatic and chemical reactions responsible for spoilage of seafood run slower at low temperatures compared to at ambient temperatures. Changes in fish freshness with temperature are shown in Figure 4.1.

4.2.1 Chilling methods of fish: traditional and advanced
4.2.1.1 Chilling with ice
For every drop in temperature of 5°C, the rate of spoilage is halved, which means that the shelf life of the fish is doubled. It is not possible to keep unfrozen fish at a temperature low enough to stop bacterial action completely, because fish begin to freeze at about −1°C. However, it is desirable to keep the temperature of unfrozen fish as close to that level as possible in order to reduce spoilage; the easiest and best way of doing this is to use plenty of ice which, when made from clean fresh

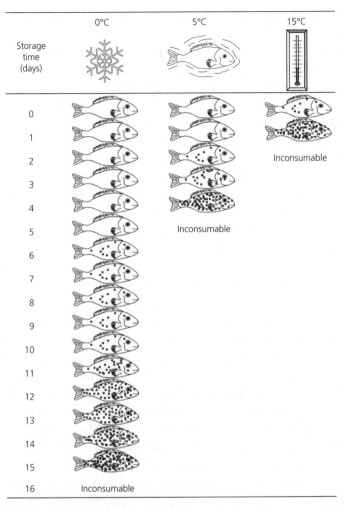

Figure 4.1 Changes in fish freshness with temperature.

water, melts at 0°C. Provided the fish is completely surrounded by melting ice, it will be chilled to about 0°C. In fact fish temperatures slightly below 0°C have been recorded in many instances in trawlers because the melting point of the ice is reduced slightly by the salts, blood and slime present in the catch. Using ice alone however, the fish temperature cannot be reduced to the point at which freezing begins, between −1°C and −2°C. According to the best technique, therefore, each fish should be in intimate contact with melting ice so that fish temperature is reduced as quickly as possible and maintained as low as possible (Merritt 1969).

Ice as a cooling medium for fish has a great deal in its favour; it has a very large cooling capacity for a given weight or volume, it is harmless, portable and relatively cheap. Ice is a portable cooling system and can be moved around from one place to another. It is especially valuable for chilling fish, since rapid cooling is possible. When fish are being cooled with ice, heat transfer is achieved by direct contact with the ice, by conduction through adjacent fish and by melt-water flowing over the fish. Cold melt-water takes up heat from the fish and when it flows over ice again it is re-cooled. Thus, intimate mixing of fish and ice not only reduces the thickness of the layer of fish to be cooled but also promotes this convective cooling interaction between ice melt-water and fish (Graham et al. 1992). Since ice will absorb large amounts of heat when it melts, it is said to have a large cooling capacity and so it can cool fish rapidly, if used correctly (Warren 1986).

Ice is its own thermostat and, since fish are mainly water, ice maintains fish at a temperature just slightly above the point at which they would begin to freeze; the point of equilibrium for sea fish, iced soon after catching, is near to –0.5°C, since the mixture usually includes some salt and blood (Graham et al. 1992). Just as water freezes at 0°C, ice will melt at a constant temperature of 0°C when heat is added to it. Thus, melting ice, if present in sufficient amounts, will maintain fish at a temperature of 0°C, but fish in melting ice will never freeze. This is important since freezing can affect the quality of the fish, unless it is carried out under very carefully controlled conditions (Warren 1986).

Historical evidence proves that the ancient Chinese utilized natural ice to preserve fish more than 3000 years ago. Natural ice mixed with seaweed was also used by the ancient Romans to keep fish fresh. However, it was the development of mechanical refrigeration that made ice readily available for use in fish preservation.

Ice is utilized in fish preservation for two reasons: (1) *Temperature reduction*. By reducing temperature to about 0°C the growth of spoilage and pathogenic microorganisms is reduced, thus reducing the spoilage rate and reducing or eliminating some safety risks. Temperature reduction also reduces the rate of enzymatic reactions, in particular those linked to early post-mortem changes extending, if properly applied, the rigor mortis period. (2) *Melting ice keeps fish moist*. This action mainly prevents surface dehydration and reduces weight losses. Melting water also increases the heat transport between fish and ice surfaces. (Huss 1995). Lower microbial load was observed in tilapia and catfish stored in ice for 7 days compared to refrigerated samples (Mohammed & Hamid 2011).

Ice has some advantages when compared with other cooling methods, including refrigeration by air. The properties can be listed as follows:

- *Ice has a large cooling capacity.* The latent heat of fusion of ice is about 80 kcal/kg. This means that a comparatively small amount of ice will be needed to cool 1 kg of fish. For example, for 1 kg of lean fish at 25°C, about 0.25 kg of melted ice will be needed to reduce its temperature to 0°C. The reason why more ice is needed in practice is mainly because ice melting should compensate for thermal losses.
- *Ice melting is a self-contained temperature control system.* Ice melting is a change in the physical state of ice (from solid to liquid), and in current conditions it occurs at a constant temperature (0°C). This is a very fortunate property without which it would be impossible to put fresh fish of uniform quality on the market. Ice that melts around a fish has this property on all contact points.

Clean ice is the easiest and best way to chill fish. The advantages of using clean ice to chill fish are:

- It is efficient at chilling fish quickly
- It is usually fairly cheap
- It is harmless (as long as it is made from clean water)
- It keeps the fish moist, shiny and attractive
- It is easily transportable from the place where it is made to the vessel.

There are various factors affecting effectiveness of icing.

Source of ice

There is often argument about whether ice made at one port is better than another, whether natural ice is better than artificial ice, freshwater ice is better than seawater ice, or new ice is better than old ice. The differences between freshwater ices of different origin are so small, that they are of no significance to those using ice for chilling fish. Ice made from tap water has the same cooling power as ice made from distilled water, and ice 3 months old is as effective as newly made ice. Ice made from hard water has the same cooling properties as ice made from soft water, although particles of ice made from hard water sometimes tend to stick together more during melting than pieces made from soft water. The effectiveness of seawater ice, in comparison with freshwater ice, is a little more in dispute. Depending on the method of manufacture, seawater ice may be less homogeneous than freshwater ice when newly made. Brine will also leach out of seawater ice during storage, so that the ice does not have a precisely fixed melting point. For this reason, fish kept in seawater ice may sometimes be at too low a temperature and become partially frozen, or the fish may possibly take up some salt from the ice. Manufacture of seawater ice is of particular advantage on board ship for augmenting port supplies on a long voyage, or in coastal communities where fresh water is so scarce and expensive that to make ice from it would be prohibitive. It is important to remember, however, that seawater for making ice must be uncontaminated; all too often the quality of coastal or harbour water makes it unsafe for use with food (Graham *et al.* 1992).

When considering the manufacture of ice on board fishing vessels, seawater will be the natural choice of raw material. When considering whether to use fresh or seawater in land-based plants, the decision will depend on several factors, such as the availability of regular supplies, the location of the ice plant and the intended use of the ice (e.g. for use on board fishing vessels or on shore). Whatever type of water is used, it must be remembered that the resultant ice will come into direct contact with food. For this reason it is essential that the water used is free from contamination that could cause risks to human health or tainting of the fish, so that it becomes unacceptable. This implies that the water must be of drinking-water quality and comply with the safety standards laid down by such bodies as the World Health Organization (Shawyer & Pizalli 2003). When seawater is frozen slowly, freshwater ice-crystals are initially frozen out of the mixture. The whole solution will not be frozen until the temperature has reached −22°C, the eutectic point. (The eutectic point is a physical constant for a mixture of given substances.) At higher freezing rates, the ice-crystals will be salt-contaminated from the very beginning but this salt will eventually migrate to the outer surface and separate during storage. As the crystals are made mainly of fresh water, the residual liquid will contain an ever increasing concentration of salt, as the temperature is reduced. The special structure of seawater ice gives it different properties from freshwater ice. Seawater ice is rather soft and flexible and, at normal sub-cooled ice temperatures of −5°C to −10°C, it will not keep the form of flakes; in fact, at −5°C, seawater ice will look rather wet. For this reason, seawater ice is usually produced at lower temperatures than freshwater ice, and often this adjustment has to be made to the ice maker. Otherwise the plant required is basically the same. Some difficulties have also been experienced with the pneumatic transportation of seawater ice. Even when sub-cooled, the conveyor raises the temperature sufficiently to make the ice soft, sticky and difficult to move.

Type of ice

This is precisely the reason that the quality of fish is preserved longer when there is an adequate amount of ice, and it is well dispersed among the fish. When the ice is in small particles, such as flakes, it does a more effective job of cooling than when it is in large pieces. Smaller ice particles give greater contact between fish and ice, and the rate of heat removal depends on the size of the contact area. Another advantage of small ice particles is that it avoids damaging fish in the lower part of the pens. Large pieces of ice can exert point forces (from the pressure developed in the lower part of the pens when they are filled) and thereby damage fish (Ronsivalli & Baker 1981). Different types of ices are produced using different ice makers. These are as follows:

Block ice: The traditional block ice maker forms the ice in cans, which are submerged in a tank containing circulating sodium or calcium chloride brine. The dimensions of the can and the temperature of the brine are usually selected to give a freezing period of between 8 and 24 hours. Too rapid freezing results in brittle ice. The block weight can vary from 12 to 150 kg, depending on requirements. The thicker the block results in a longer freezing time. The size of the tank required is related to the daily production. A travelling crane lifts a row of cans and transports them to a thawing tank at the end of the freezing tank, where they are submerged in water to release the ice from the moulds. The cans are tipped to remove the blocks, refilled with fresh water and replaced in the brine tank for a further cycle. Block ice plants require a good deal of space and labour for handling the ice. The latter factor has been the main reason for the development of modern automatic ice-making equipment. The block ice can be easily stored in any insulated or refrigerated shore facility. It has a density of about 720 kg/m³. Block ice has some advantages and disadvantages, over other forms of ice.

Advantages:

- Storage, handling and transportation are simple and easy and can all be simplified if the ice is in form of large blocks.
- The melting rate is relatively slow, and therefore losses during storage and distribution are minimal.
- The plant is robust engineering and relatively simple to maintain by a competent mechanical engineer.
- For the same container, more block ice (in weight) can be packed than with the other types of ice.
- The ice is compact and therefore less storage space is required.
- With an appropriate ice-crushing machine block ice can be reduced to any particle size but the uniformity of size will not be as good as that achieved with some other forms of ice. In some situations, block ice may also be reduced in size by a manual crushing method.

Disadvantages:

- Before it can be used with fish, block ice must be broken up to make crushed ice; a long time period is required (8–36 h) to complete the freezing of water in cans (block size from 12 to 140 kg).
- There are high labour costs and continuous attention to operations.
- It is not a continuous automatic process and it takes a long time to produce ice from first start-up.
- Care must be taken to break the ice blocks into small pieces as large pieces would not wrap well around fish in a container, and would result in air gaps around the fish leading to poor chilling.
- Space requirements for the ice plant itself are greater than for modern automatic ice-makers.
- Adequately treated brines are necessary to minimize equipment corrosion; ice must be crushed before use.

Block ice must be crushed before loading on the fishing vessel. Crushed ice, because of possible sharp edges of the pieces may not be best suited for making good contact with the fish. Sharp edges of crushed ice blocks may also cut and damage fish (Sikorski 1990; Graham *et al.* 1992; Blanc 1995; Shawyer & Pizalli 2003).

Containerized block ice plants are available that house the ice plant, ice store and complete refrigeration and electrical systems inside standard containers. This allows portability, ease of transport by sea and land, better reliability, and significantly shorter installation and break in periods than traditional no-containerized types. These advantages are important, particularly in remote areas where there is limited refrigeration and maintenance expertise. These units are fitted into standard 13 m (40 ft) containers, and are easy to install. They only require a levelled foundation and to be under cover for protection against the weather, and they can be built in tropical climates and coastal conditions. Units are available that produce blocks of various sizes from 12.5 to 25 kg (Shawyer & Pizalli 2003).

Flake-ice: Flake-ice can be defined as dry and sub-cooled ice in small flat pieces with an irregular wafer shape 100 to 1000 mm^2 in area. Flake-ice is produced by freezing water on the surface of a refrigerated drum and scraping the ice off with a blade of usually 2 to 3 mm thickness. It has the largest surface area per unit weight and a density of 480 g/m^3 (Sikorski 1990; Shawyer & Pizalli 2003). The water freezes on the surface and forms thin layers of ice. A scraper removes the sub-cooled ice, which breaks into small pieces resembling splinters of glass. These pieces of ice usually fall from the drum directly into a refrigerated compartment for storage. The cooled cylinder can rotate either in a vertical or horizontal plane (Shawyer & Pizalli 2003). In some models, the cylinder or drum rotates and the scraper on the outer surface remains stationary. In others, the scraper rotates and removes the ice from the surface of a stationary drum, in this case, built in the form of a double-walled cylinder. It is usual for the drum to rotate in a vertical plane but in some models the drum rotates in a horizontal plane. One distinct advantage of the rotating drum method is that the ice-forming surfaces and the ice-release mechanism are exposed and the operator can observe whether the plant is operating satisfactorily. The machine with the stationary drum has the advantage that it does not require a rotating seal on the refrigerant supply and takeaway pipes. However, this seal has been developed to a high degree of reliability in modern machines. The ice is sub-cooled when harvested; the degree of sub-cooling depends on a number of factors, but mainly the temperature of the refrigerant and the time allowed for the ice to reach this sub-cooled temperature. The sub-cooling region of the drum is immediately before the scraper where no water is added for a part of the drum's rotation and the ice is reduced in temperature. This ensures that only dry sub-cooled ice falls into the storage space immediately below the scraper.

The refrigerant temperature, degree of sub-cooling and speed of rotation of the drum are all variable with this type of machine and they affect both the capacity of the machine and the thickness of the ice produced (Blanc 1995). A variation on flake-ice is known as chip ice. Chip ice is manufactured by flowing water inside the ice-making cylinder, which is surrounded by an evaporating coil. The water is frozen inside the cylinder at an evaporator temperature of −12 to −30°C and removed with an auger revolving inside the cylinder and pushing the ice upwards. In the upper part of the cylinder the ice is pressed, frozen further and ejected through the top of the cylinder. Chip ice has a temperature of −0.5°C and an average thickness of 7–8 mm (Shawyer & Pizalli 2003).

Flake-ice has some advantages and disadvantages.

Advantages:

- It wraps very closely around fish, giving very rapid chilling. It is easy to handle. Thus, it can be very evenly distributed between the layers of fish.
- It is easy to store and handle when adequately designed sub-cooled (−5°C) insulated storage is provided.
- Due to the fact that flake-ice is slightly sub-cooled (−5°C to −7°C), it can give off 83 kcal/kg when melting from ice to water; therefore slightly more heat can be extracted than with other types of ice at a temperature of 0°C (80 kcal/kg).
- Flakes present the maximum cooling surface for a given amount of ice. Flake-ice has a larger heat-exchange surface than most other types of ice, therefore heat transfer between fish and ice occurs faster and more efficiently.
- The plant is small and compact, using less space than block ice plants.
- The manufacture of ice begins within a very short time of starting the machine, almost allowing 'ice on demand'.
- Ice is ready to use immediately after manufacture (does not need crushing).

Disadvantages:

- Flake-ice melts quicker than the other types of ice; therefore it is not so good for transporting over long distance to where it is needed for chilling of fish.
- The plant is less robust and more complex and requires skilled engineers for maintenance.

- Weight for weight, flake-ice requires more storage space.
- The ice produced has to be weighed before sale rather than being sold by the unit (Blanc 1995; Sawyer & Pizalli 2003).

Tube ice: Tube ice is formed on the inner surface of vertical tubes and is produced in the form of small hollow of cylinders of about 50 × 50 mm with a wall thickness of 10 to 12 mm. The tube ice plant arrangement is similar to a shell and tube condenser with the water on the inside of the tubes and the refrigerant filling the space between the tubes. The machine is operated automatically on a time cycle and the tubes of ice are released by a hot gas defrost process. As the ice drops from the tubes a cutter chops the ice into suitable lengths, nominally 50 mm, but this is adjustable. Transport of the ice to the storage area is usually automatic, thus, as in the flake-ice plant, the harvesting and storage operations require no manual effort or operator attendance. Tube ice is usually stored in the form it is harvested, but the particle size is rather large and unsuitable for use with fish. The discharge system from the plant therefore incorporates an ice crusher, which can be adjusted to give an ice particle size to suit the customer's requirement. The usual operating temperature of this type of plant is –8°C to –10°C. The ice will not always be sub-cooled on entering the store, but it is usually possible to maintain the store at –5°C since the particle size and shape allow the ice to be readily broken up for discharge.

Plate ice: Plate ice is formed on one face of a refrigerated vertical plate and released by running water on the other face to defrost it. Other types form ice on both surfaces and use an internal defrost procedure. Multiple plate units are arranged to form the ice-making machine and often these are self-contained units incorporating the refrigeration machinery in the space below the ice-maker. The optimum ice thickness is usually 10 to 12 mm and the particle size is variable. An ice breaker is required to break the ice into a suitable size for storage and use. Water for defrost requires heating if its temperature is less than about 25°C; below this value the defrost period is too long, resulting in a loss in capacity and an increase in cost. This machine, like the tube ice machine, operates on an automatic timed cycle and the ice is conveyed to the storage area, or, if the machine can be located directly above the storage space, harvesting can be achieved using gravity flow.

Slush ice: The cooling unit for making 'slush' ice is called a scraped-surface heat exchanger. It consists of concentric tubes with refrigerant flowing between them and water in the inner tube. The inner surface of the inner tube is scraped using, for example, a rotary screw. The small ice-crystals formed on the tube surface are scraped off and mixed with unfrozen water. This results in slurry of ice and water, which may contain up to 30% water by weight. This mixture may be pumped or, after removing most of the water in a mechanical separator, used as a 'dry' form of ice.

When mixed with water, the crystals allow slurry to be pumped easily by flexible hoses to wherever it is required on the boat. This ice acts in a similar manner to chilled sea water (CSW) when in slurry form, and as such can be used in CSW tanks or fish holds. In slightly less liquid form it can also be used to bulk pack fish in tote boxes. Slush ice is a mixture of ice-crystals in water and water slurry. The ice is formed by freezing ice-crystals out of a weak brine solution in a tube-in-tube heat exchanger, also called a scraped-surface heat exchanger. Water is frozen as tiny round/ellipsoid crystals (about 0.2 to 1.3 mm diameter) on the inner-tube surface and a rotary screw conveyor moves the ice-crystals out of the heat exchanger into a storage tank with water. The resulting mixture of ice and water (slush ice) can be pumped from the storage tanks through piping or hoses to the fish-chilling area or directly to an insulated container. The density and fluidity of slush ice can be adjusted by regulating the amount of water added; so that they can be tailored to different applications. The advantages claimed for slush ice for chilling fish are as follows:

- It ensures faster and even chilling of fish to or below 0°C, due to improved heat transfer.
- It gives better contact with fish surface without bruises or pressure damage.
- It is claimed that ice contamination is significantly reduced due to the sealed system design of the ice-maker and storage.
- Ice can be pumped directly to where it is needed so there is not necessarily a need for storage (Blanc 1995; Sawyer & Pizalli 2003).

Properties of ice

In order to understand why ice is so useful for chilling fish, it is first necessary to consider the nature and properties of ice. When water freezes at a temperature of 0°C it experiences a phase change from a liquid to a solid, familiar to all as ice. A quantity of heat has to be removed from the water to turn it into ice, and the same amount of heat has to be added to melt it again. The temperature of a mixture of water and ice will not rise above 0°C until all the ice has melted. A given amount of ice always requires the same amount of heat to melt it; 1 kg of ice needs 80 kcal to change it into water, thus the latent heat of fusion of ice is 80 kcal/kg. This amount of heat is always the same for ice made from pure water, and is very little different for ice made from fresh water from almost any commercial source. Ice needs a large amount of heat to melt – it has a large reserve of 'cold' – and this is one reason why it is widely used in the fishing industry as a means of chilling fish.

Quantity of ice

It is possible to calculate the ice requirement if the operational conditions are known. These conditions are often variable and un-predictive. Only a series of tests, under operational conditions, will establish the correct fish to ice ratio to be used to cool the fish and maintain chilled temperatures during the entire storage period. Calculated values of ice usage can provide valuable information at the planning and design stages, and also promote a better understanding of the relative effect of the various elements that influence the rate of ice melting. In addition, by considering all possibilities and calculating ice requirements, a more rational judgement can be made when selecting equipment and procedures to be used. To determine the ice requirement, it is necessary to calculate the quantity of ice to cool the fish, and also the quantity of ice required to maintain the fish at a chill temperature throughout the storage period. In addition, allowance has to be made to allow for losses and other contingencies in order to determine the ice-manufacturing requirement (Graham *et al.* 1992).

Calculation of the ice requirement for cooling fish

The quantity of ice needed to cool fish depends on the weight of fish to be cooled, the temperature of fish at the start of chilling, the length of time the fish are required to be kept chilled, and how much the fish and ice protected from outside heat sources. It is possible to calculate how much ice is required to chill fish, but in practice a rough guide is to use at least one part ice to one part fish by weight for the initial chilling. Extra ice can then be added as needed. It is good practice to always have some ice present at all stages of storage and distribution (Warren 1986). The mass of ice needed to cool fish from the initial temperature to the final holding temperature can be calculated from an expression, which equates the heat taken up by the ice, on the left side of the equation, with the heat lost by the fish, on the right side of the equation (Graham *et al.* 1992).

$$(M_i)(L_i) = (M_f) \times (C_{pf}) \times (t_s - t_c)$$

Where M_i = mass of ice that melts (kg); L_i = latent heat of fusion of ice (80 kcal/kg); M_f = mass of fish (kg); C_{pf} = specific heat of fish (kcal/kg°C); t_s = initial temperature of fish (°C); t_c = final temperature of fish (0°C). From this equation the ice requirement will therefore be:

$$M_i = (M_f) \times (C_{pf})(t_s - t_c)/L_i$$

For larger volumes of fish, the amount of ice should be multiplied by the actual weight of fish to be stowed in the hold. For example, at an ambient temperature of 30°C, a load of 1000 kg of fish would require 340 kg of ice, or slightly over a ratio of 1:3 ice to fish just to cool the fish to 0°C. This does not allow for losses due to heat infiltration or extra ice necessary to maintain the fish at 0°C for the rest of the trip. In practice, therefore, much greater quantities are required to ensure that the fish remains chilled once its temperature has been reduced to 0°C, and that it can be stored at chill temperature for some time. It is a generally accepted 'rule of thumb minimum' to use an ice to fish ratio of 1:1 in the tropics. In many instances ratios of up to 3:1 ice to fish are used. The main influence on the ratio is the length of the fishing trip (Shawyer & Pizalli 2003).

Calculation of the ice requirement for the storage of fish

Even if you are concerned with only one batch of fish held in identical containers, there are likely to be variations in ice melting rates, which make it difficult to calculate the ice requirement accurately. If, for instance, the containers are stacked, then ice melting may be different in containers located at the top, bottom, sides and within the stack. In spite of the obvious difficulties and likely inaccuracies, a calculation of the ice melting rate can still be useful at a planning stage to enable comparisons to be made between different options, and to allow preliminary estimates of quantities, costs and equipment to be made.

It would be difficult to identify containers which would eventually be located at more favourable locations within a stack. Therefore, all containers should be treated equally and the assumption made that each container is fully exposed to the surrounding air. As a first step, heat transfer may be calculated using the following simple expression:

$$q = A \times U(t_o - t_c) \, \text{kcal/day}$$

Where: q = heat entering the container (kcal/day); A = surface area of the container (m^2); U = overall heat transfer coefficient kcal/day m^2 °C); t_o = temperature outside the container (°C); t_c = temperature inside the container (0°C).

This overall calculation of heat transfer may require to be done in parts if, for instance, the lid or base of the container is made of a different material or has a different thickness. The heat transfer through the various areas is then added together to give the total heat transfer. The heat entering melts the ice, therefore it follows that:

$$q = L_i \times m_i \, \text{kcal/day}$$

Where: q = heat required to melt ice (kcal/day); L_i = latent heat of fusion of ice (usually taken as 80 kcal/kg); m_i = mass of ice melted (kg/day).

In order to develop a mathematical expression for ice melting rate during the storage period we suppose that ice melting inside the containers is only due to heat transferred from the surrounding air. In this steady state approach, quantities should be equal, therefore it follows that:

$$L_i \times m_i = A \times U \times (t_o - t_c)$$

Thus, the ice requirement will be:

$M_i = A \times U \times (t_o - t_c)/L_i$ kg/day. Table 4.1 shows the theoretical amount of ice needed according to various temperatures.

If the fish containers are exposed to direct sunlight during the storage period, the above calculation, which is only based on the conductance of heat due to the difference between the internal and external temperatures, will result in an underestimation of the ice requirement. To include the element of ice melting due to radiated heat will make the calculation extremely difficult. Therefore, if the containers cannot be protected from direct sunlight or any other radiated heat source during the storage

Table 4.1 Theoretical weight of ice needed to chill 10 kg of
fish to 0°C from various ambient temperatures

Temperature of fish (°C)	Weight of ice needed (kg)
30	3.4
25	2.8
20	2.3
15	1.7
10	1.2
5	0.6

Food and Agriculture Organization of the United Nations 1984
Eds J Brox *et al.* Planning and engineering data 4. Containers for
fish handling. Reproduced with permission.

period, the calculated values for the ice requirement should be upgraded,
or used with caution (Graham *et al.* 1992).

The amount of ice needed to keep fish fresh is economically more
important in tropical countries, since the warmer climate means that ice
melting rates are higher. The ice required for cooling the fish from the
initial temperature is fixed and cannot be reduced, but the use of
insulation and refrigeration can considerably reduce the ice requirement
during the subsequent storage period. Factors other than higher ambient
temperatures can result in an increase in the ice requirement in tropical
countries. The collection and marketing system may require that the fish
and ice be separated for check weighing and sorting and, if the correct
procedure is followed, the old ice should be discarded and new ice used
for re-icing. It is also advisable in tropical countries to pre-cool water used
during processing in order to avoid undesirable rises in fish temperature
which would accelerate fish spoilage. Keeping the fish chilled at this stage
also avoids the need to re-cool later. In more sophisticated systems, water
pre-cooling can be achieved using a mechanical refrigeration system and
heat exchanger, but a more simple method is to merely add ice to the
water in the supply tank (Graham *et al.* 1992).

Insulation

It is important to know how heat is transferred in fish holds. Heat is
transferred by conduction, convection or radiation, or by a combination
of all three. Heat always moves from warmer to colder areas; it seeks
a balance. If the interior of an insulated fish hold is colder than

the outside air, the fish hold draws heat from the outside. The greater the temperature difference, the faster the heat flows to the colder area.

- *Conduction*: By this mode, heat energy is passed through a solid, liquid or gas from molecule to molecule. In order for the heat to be conducted, there should be physical contact between particles and some temperature difference. Therefore, thermal conductivity is the measure of the speed of heat flow passed from particle to particle. The rate of heat flow through a specific material will be influenced by the difference of temperature and by its thermal conductivity.

- *Convection*: By this mode, heat is transferred when a heated air/gas or liquid moves from one place to another, carrying its heat with it. The rate of heat flow will depend on the temperature of the moving gas or liquid and on its rate of flow.

- *Radiation*: Heat energy is transmitted in the form of light, as infrared radiation or another form of electromagnetic waves. This energy emanates from a hot body and can travel freely only through completely transparent media. The atmosphere, glass and translucent materials pass a significant amount of radiant heat, which can be absorbed when it falls on a surface. It is a well-known fact that light-coloured or shiny surfaces reflect more radiant heat than black or dark surfaces; therefore the former will be heated more slowly.

In practice, the entry of heat into fish holds/fish containers is the result of a mixture of the three modes mentioned above, but the most significant mode is by conduction through walls and flooring.

The primary function of thermal insulation materials used in small fishing vessels using ice is to reduce the transmission of heat through fish hold walls, hatches, pipes or stanchions into the place where chilled fish or ice is being stored. By reducing the amount of heat leakage, the amount of ice that melts can be reduced and so the efficiency of the icing process can be increased. As has already been discussed, ice is used up because it removes heat energy from the fish but also from heat energy leaking through the walls of the storage container. Insulation in the walls of the container can reduce the amount of heat that enters the container and so reduce the amount of ice needed to keep the contents chilled. The main advantages of insulating the fish hold with adequate materials are:

- to prevent heat transmission entering from the surrounding warm air, the engine room and heat leaks (fish hold walls, hatches, pipes and stanchions);

- to optimize the useful capacity of the fish hold and fish-chilling operating costs;
- to help reduce energy requirements for refrigeration systems if these are used.
- Insulation can be used in various ways and the choice between one system and another will depend mainly on local conditions (Shawyer & Pizalli 2003).

Because hold space is often at a premium on small vessels and the costs of insulation can amount to a significant proportion of the costs involved in construction, the choice of insulation material can be very important. Several thermal insulation materials are used commercially for fishing vessels, but few are completely satisfactory for this purpose. The main problems are lack of sufficient mechanical strength and moisture absorption. Desired characteristics of insulation materials as follows:

Thermal conductivity: The best insulation materials should have the lowest thermal conductivity, in order to reduce the total coefficient of heat transmission. Thus, less insulating material will be required. Dry static gas is one of the best insulating materials. The insulating properties of commercially available insulating materials are determined by the amount of gas held inside the material and the number of gas pockets. Therefore, the higher the number of cells (which can maintain the gas stagnant) and the smaller their size, the lower the thermal conductivity of such insulation material. These cells should not be interlinked, as this will allow convection of heat.

Moisture-vapour permeability: The best insulation materials should have very low moisture-vapour permeability. Thus, water absorption becomes negligible and condensation and corrosion are minimized.

Resistance/installation features: The insulation material should be resistant to water, solvents and chemicals. It should be durable, and not lose its insulating efficiency quickly. It should allow a wide choice of adhesives for its installation. It should be easy to install, of light weight and easy to handle. Ordinary tools can be used for its installation. It should be economical, with significant savings on initial cost as well as savings on long term performance. It should not generate or absorb odours. It should be unaffected by fungus or mildew and should not attract vermin. It should be dimensionally stable, so it will not crumble or pack down.

Safety features: The insulation material should be rated as non-flammable and non-explosive. In the event that the insulation material burns, the products of combustion should not introduce toxic hazards.

A wide range of insulation materials is available; however, few meet the requirements of modern fish hold construction. Selection of insulation material should be based on initial cost, effectiveness, durability, the adaptation of its form/shape to that of the fish hold, and the installation methods available in each particular area. From an economic point of view, it may be better to choose an insulating material with a lower thermal conductivity rather than increase the thickness of the insulation in the hold walls. By reducing the thermal conductivity, less insulation will be required for a given amount of refrigeration and more usable volume will be available in the fish hold. The space occupied by the insulation materials in fishing vessels can represent, in many instances, about 10 to 15% of the gross capacity of the fish hold.

Polyurethane foam: One of the best commercially available choices of insulation material for fishing vessels is polyurethane foam. It has good thermal insulating properties, low moisture-vapour permeability, and high resistance to water absorption, relatively high mechanical strength and low density. In addition, it is relatively easy and economical to install. The main ways polyurethane foams can be applied and used are as rigid boards/slabs and pre-formed pipes, which can be manufactured in various shapes and sizes. The main applications of these types of foam are in chill rooms, ice stores and cold stores. Structural sandwich panels incorporating slabs of foam can be produced for prefabricated refrigerated stores.

Expanded polystyrene: Through polymerization styrene can be made into white pearls/beads of polystyrene plastic. These beads can then be expanded to form foam known as expanded polystyrene. There are two main ways of making of expanded polystyrene: by extrusion and by moulding of slabs. Extruded foams are made by mixing the polystyrene with a solvent, adding a gas under pressure and finally extruding the mixture to the required thickness. The extrusion process improves the characteristics of the final foam, such as its mechanical resistance, producing non-interconnecting pores and a more homogeneous material. The mechanical resistance of expanded polystyrene foams can vary from 0.4 to 1.1 kg/cm^2. There are several grades of foams available with densities from 10 to 33 kg/m^3, with thermal conductivities that are lower with the increase in density

Expanded perlite: Perlite is a volcanic rock containing from 2 to 5% bonded water. It is a chemically inert substance composed basically of silica and aluminium, but some impurities, such as Na$_2$O, CaO, MgO and K$_2$O, which are hygroscopic, can absorb moisture easily. Therefore, depending on the storage conditions and the quality of the perlite, moisture absorption can be minimized. The average density of expanded perlite is about 130 kg/m^3 and its thermal conductivity is about 0.04 kcal/m/ h/ °C (0.047 W/m/°C). The perlite is expanded by means of rapid heating at a temperature between 800 and 1200°C. The vaporization of the bonded water and the formation of natural glass results in the expansion of the perlite particles, which have a granular shape.

Fibreglass: Fibreglass matting is also used as insulating material. The advantages of fibreglass are high resistance to fire, microbiological attack, most chemicals and heat. Fibreglass insulation is available in rolls of different thickness, also called blankets and mats. The width of the blankets and mats will depend on the way they are to be installed and some come faced on one side with foil or Kraft paper, which serve as vapour barriers.

Many types of containers, constructed from a variety of materials, are used for the transport of ice and fish – from simple baskets of woven reeds, bamboo, cane or grasses, to containers made from wood, metals and plastics. Although estimates vary, in some situations a high proportion of fresh fish caught in tropical and subtropical areas may be wasted, with the major loss in quality and value occurring between harvesting operations and first sale in landing areas. It is envisaged that with the increased availability and wider use of properly designed containers for use on canoes and small fishing vessels, there will be scope for reducing wastage of fresh fish in small-scale fisheries. The main functions of an insulated fish container on board canoes and small fishing vessels are:

- to make handling easier (by reducing the handling frequency of individual fish) and protect the fish from the risk of physical damage;
- to maintain fish quality, by ensuring adequate chilling and low ice-melting rates as a result of reduced heat infiltration through container walls;

- to improve fish-handling practices and so lead to better quality fish being landed, making longer fishing trips and better fish prices possible for fishermen.

The effectiveness of insulated containers in reducing ice melting is an important criterion in the evaluation and selection of such containers. It is more likely that the advantages that insulated containers offer will be fully appreciated by small-scale fishermen in tropical climates where ice melting rates are much higher than in cold or temperate climates.

Insulated covers: These are made to fit snugly over a stack of ordinary plastic fish boxes to maximise draught proofing whilst providing heat insulation from the sides and top. Being flexible they can be folded and stored when not in use.

Insulated boxes: By completely surrounding the fish and ice with insulation, insulated boxes offer a better degree of heat insulation than covers. Being portable they can be used to keep ice both at sea and during transport from the ice plant to the vessel. They may also be used to hold fish and ice after landing.

Insulated containers: These are larger equivalents of insulated fish boxes and are versatile as ice keeps from which to ice fish under cover or hold iced fish. This may be in boxes or loose. Using containers with the drainage bungs enables slush icing – a mixture of seawater, ice and melt water – to rapidly cool the fish and avoid pressure effects. These containers are less portable than the insulated boxes and would tend to take up a more or less 'permanent' position on the deck of the fishing boat.

Icing procedures

For efficient and rapid cooling, ice must be used correctly. It is important to completely surround the fish with ice so that it is in full contact with the product. For this, ice needs to be in very small pieces or in slush form, or perhaps made into ice-water slurry as in CSW applications. A few large lumps of ice scattered over the fish do not have the same cooling effect as small particle ice packed around the fish. Whether in boxes or stowed in bulk in fish-hold pens, fish also need to be carefully layered in ice. The proper icing procedure is shown in Figure 4.2.

Fish and ice must be packed together in the ice box so that each fish is completely surrounded by ice. The fish should not be touching each

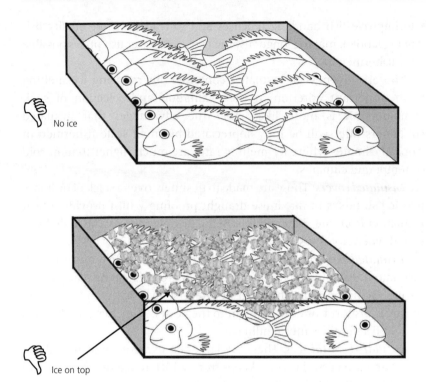

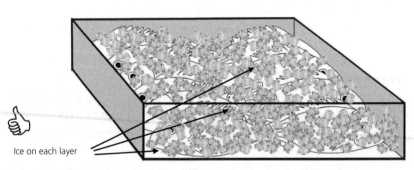

Figure 4.2 Icing of fish in boxes.

other or the side of the container. The first step is to put a layer of ice into the container followed by a layer of fish. Fish, whether it has been gilled and gutted or not, should be placed belly down. The fish are then covered by a layer of ice, and a second layer of fish is added. Cover again with ice and repeat the process until the container is full. As a general rule, to properly chill fish to 0°C, a ratio of one part ice to one part fish is needed (Blanc 1995).

For optimal use of ice, the following points should be taken into account:

- All ice used must be clean and of small particle size for maximum contact. Block ice must be finely crushed to prevent large particles from damaging the fish.
- The proper ratios of fish to ice must be observed. In temperate climates, one part fish to one part ice is common. In tropical conditions, one part fish to three parts ice is not unusual.
- Areas of heat penetration into the hold, such as the engine room bulkhead and hull sides, must be given extra layers of ice to compensate for rapid ice loss in these areas, particularly if insulation is poor.
- The last layers of fish near the deck head should have extra layers of ice to fully cover the fish and allow for any extra melting from heat penetration through the deck.
- Fish and ice must be carefully and evenly stowed to allow even distribution of both. Shelves and boxes must not be overfilled or crushing damage to the fish will result.
- Fish temperatures at the dockside when discharging should be between 0°C and 2°C, and there should also be considerable amounts of ice still evenly distributed among the fish.
- Ice must be layered under, around and on top of the fish.

4.2.1.2 Chilling with water–ice mix/slurry ice

Another effective method for chilling fish is to use sea-water/ice slurry. To make the slurry, mix two parts of ice to one part of sea water. Only crushed ice or flake-ice are used.

Advantages:

- The fish can be dropped directly in the slurry after it is caught, instead of taking much longer to properly pack the fish in ice. This saves time, particularly when fish are biting.
- Chilling is very fast because there is 100% contact between the fish and the chilling medium of around 1°C (Graham *et al.* 1992).

Slurry ice, also known as fluid ice, slush ice, liquid ice, flow ice or binary ice, has been reported to be a promising technique for the preservation of aquatic food products in an ice-water suspension at sub-zero temperatures. It has shown to provide several advantages towards flake-ice such as: lower temperature, faster chilling, lower physical damage to product and better heat exchange power (Losada *et al.* 2004a). Slurry ice is a mixture of fine ice-crystals, water and freezing point depressant additives. The typical ice-crystal size ranges between 0.1 to 1 mm in diameter. The size depends on additive type and concentration and type of ice generator (Hagg 2005). Ice slurries are normally ice-crystals distributed in water or an aqueous solution. The Different substances are added. The purposes of fading different substances are freezing point depression for applications below 0°C, decreasing the viscosity, increasing the thermal conductivity of the fluid phase, reduction of corrosive behaviour of the ice slurry, prevention of agglomeration (Egolf & Kauffeld 2005). Using salted water as raw material for production, slurry ice is composed of millions of microscopic spherical ice-crystals suspended in seawater or brine. These structural characteristics provide the ice with a superior ability to chill fish due to its better heat exchange power. The physical characteristics of this suspension provide slurry ice with its ability to provide rapid chilling of fish and lower fish temperature (Huidobro *et al.* 2001). Snow crystals mixed with water lead to slurry, which occurs in nature and is known to everybody who has walked in winter in rainy weather in a snowy countryside. Another kind of ice slurry is created in the growth of falling hailstones, when the heat transfer is insufficient to freeze the accreted super cooled cloud droplets (Egolf & Kauffeld 2005).

Ice could be produced in different forms such as block, cube and tube or flake-ice. Most of these forms of ice require a certain degree of manual operation for transportation from one place to another, and have rather sharp edges that may damage a product's surface when used for direct contact chilling. Furthermore, they are usually quite coarse and have poor heat transfer performance when releasing their latent heat of fusion. Slurry ice refers to a homogeneous mixture of small ice particles and carrier liquid. The liquid can be either pure freshwater or a binary solution consisting of water and a freezing point depressant. Sodium chloride, ethanol, ethylene glycol and propylene glycol are the four most commonly used freezing point depressants in industry. Over the last two decades interest in using phase-change ice slurry coolants has grown significantly. Relevant characteristics of slurry ice are: (1) its faster chilling rate, which is a consequence of its higher heat-exchange capacity; (2) the reduced physical damage caused to food products by its microscopic spherical particles as compared with the damage elicited by flake-ice; (3) the sub-zero storage temperature of slurry ice; and (4) the prevention of dehydration events due to the full coverage of the fish surface (Losada *et al.* 2004b, 2007). The main advantage of slurry ice is its pumpability. Slurry ice can be pumped through pipes and is storable in all type of tanks and containers (Huidobro *et al.* 2001). From an economical point of view, slurry ice is pumpable up to an ice fraction of about 30% (Hagg 2005). On the other hand, the absence of air pockets delays the biochemical processes related to the presence of oxygen (Huidobro *et al.* 2002).

Slurry ice has a high energy storage density because of the latent heat of fusion of its ice-crystals. It also has a fast cooling rate due to the large heat transfer surface area created by its numerous particles. The slurry maintains a constant low temperature level during the cooling process, and provides a higher heat transfer coefficient than water or other single phase liquids. These features of ice slurry make it beneficial in many applications. For example, the ice-slurry based thermal storage system produces and stores cold in the form of a dense ice slurry during night time hours when electricity is cheap, and the cold energy can then be quickly released by melting the ice slurry for air-conditioning of buildings during daytime hours when electricity might be several times more expensive (Kauffeld *et al.* 2005, 2010).

There are several ice making methods that manifest themselves as commercial ice-making devices; the device selected for a given ice-slurry cooling application must be chosen carefully because of the differences in the type of ice particles they produce. Most ice-slurry installations use ice generators of the scraped surface type. In many of the installations, initial investment costs are higher due to this kind of ice generator. Often, operating costs are similar to operating costs with other refrigeration systems.

Ice slurry as a new technology to maximize the chilling speed of fish has received great attentions over the past 25 years. After almost 30 years of continuous efforts from the manufacturers and research organizations, ice slurry is now well recognized, not only as an incomparable cooling technology, but also as an excellent preservation medium. Ice slurry is increasingly used for chilling, storage and transportation of fish on board fishing vessels and barges, at farms, and inside processing plants. Success has been reported for almost all major fish species such as tuna, yellowtail, salmon, cod, haddock, hake, herring, mackerel, sardine, shrimp, mussels and lobster, and turbot (Kauffeld *et al.* 2010; Rodriguez *et al.* 2004; Aubourg *et al.* 2007; Rodriguez *et al.* 2006; Erikson *et al.* 2011). It was reported that good results were found for albacore tuna stored on board with slurry ice (Price *et al.* 1991). In another study, inhibitory effect of slurry ice on chemical mechanisms (nucleotide degradation, formaldehyde production, interaction compound formation, and electrophoretic sarcoplasmic protein profiles related to quality loss of European hake has been reported (Losada *et al.* 2004b). Slurry ice was used to chill deepwater pink shrimp and it was observed that slurry ice improved the quality of fish compared to flake-ice (Huidobro *et al.* 2002). On the other hand lipid hydrolysis and oxidation were inhibited during the storage in slurry ice of horse mackerel (Losada *et al.* 2005). Preliminary slurry ice chilling increased the shelf life of frozen sardine (Losada *et al.* 2007). Slurry ice application extended the shelf life of ray (*Raja clavata*) (Mugica *et al.* 2008). Slurry ice treatment was checked alone and in combination with ozone and compared to traditional flake icing during the storage. Slurry ice showed an inhibitory effect on lipid damage during storage, as well as an inhibition of nucleotide autolytic degradation. Ozonized slurry ice did not provide differences ($p > 0.05$) from slurry ice alone when considering lipid hydrolysis, nucleotide degradation and some lipid oxidation indices (Losada *et al.* 2004a).

The rate at which heat can be removed from seafood during chilling depends on several parameters. In the case of packaged aquatic food products, size and shape will affect the rate of heat transfer while in the case of unpackaged products special attention should be paid to the relative humidity conditions, with a view to prevent surface dehydration of the fish material. In the case of seafood, chilling technologies are essential for each step in the production chain from the fish hook to the fork. Such steps are slaughter, storage, long-distance transport and each step involved in local distribution and domestic storage. Recently, newer chilling systems have enabled the storage at sub-zero temperatures of aquatic food products through the addition of salts and other compounds to ice-water mixtures. These are called 'slurry ice systems' (Pineiro *et al.* 2004).

4.2.1.3 Use of chilled /refrigerated sea water

An alternative method of lowering the temperature of fish is to immerse them totally in some chilled liquid medium held in tanks. The two basic systems used to hold fish in tanks are the refrigerated sea water (RSW) system, where some mechanical means is used to chill sea water down to about −1°C or the CSW system where freshwater ice is mixed with sea water to lower the temperature (Wheaton & Lawson 1985). The terms RSW and CSW describe seawater that has been cooled to just below 0°C. In some cases, brine of about the same salinity as seawater is used. There is no clear distinction between the two terms. RSW is generally used when a mechanical refrigeration unit cools the seawater and CSW is more often used when ice is added for cooling. This system is used for improve fish freshness for inshore fishery by cooling down seawater temperature between 0°C to −5°C with refrigeration system. The principal function of the chilling process is by repeated pumping up sea water in the fish hold by using circulation pump force to the sea water chiller and returning chilled water back into fish hold. These types of systems have been used successfully to hold salmon, tuna, halibut, herring and certain species of shellfish. It is reported that RSW storage temperature has a significant effect on post-mortem skin and fillet colour retention, and muscle biochemistry in Australasian snapper (*Pagrus auratus*) (Tuckey *et al.* 2012). Generally, oily species are more suitable for storage using these systems. Fish stored in RSW and CSW tanks have buoyancy almost equal to their own weight; therefore they are not subject to crushing as in boxed or bulked stowage. When fish are stored on ice it not always possible to get the fish completely surrounded by the chilling medium. However, with RSW or CSW systems it is easier to get the medium to completely surround each individual fish. In CSW, the fish are surrounded with a mixture of ice and water, thereby achieving maximum contact between fish and coolant. When enough ice is added to the system, it will bring the temperature of the water down to 0°C, and it will continue to remove the heat from the water, that the water, in turn, removes from the fish. The transfer of heat from the fish to the water and from water to ice will continue until the system is brought to a state of temperature equilibrium. Temperature uniformity is enhanced by the thorough mixing due to the action of the sea, as one might expect. There is no need to add ice at sea; however, seawater must be added to the ice as soon as possible. Otherwise, the ice tends to clump before slush can be obtained. The fish are less damaged in CSW because of the buoyant effect of the water. Also, fish can be unloaded very quickly with pumps. CSW is effective because the water component of the CSW establishes maximum contact between the cooling medium and the fish. Therefore, the cooling rate of fish in CSW is higher than that of fish in ice. Adequate mixing should be provided and thus the fish do not clump together and create the development of anaerobic conditions. The most important advantage of RSW and CSW systems over icing is the ease of storage on board and the ease of unloading, making the less laborious and avoiding additional handling of fish (SEAFDEC 2005). The other advantages over stowage in ice; the catch is cooled more rapidly, less effort is required to stow and unload it, and there is less likelihood of fish being crushed or losing weight. In addition, sea water can be safely lowered in temperature to about −1°C without freezing the fish contained in it. Effective washing and bleeding, and a tendency to firm the flesh of fish, which can aid further processing, are other advantages of chilled sea water (Kelman 1977).

In some cases where large numbers of fish are caught in a short period of time, it is often not possible to ice them properly. Where vessels employ chilling tanks the loading and unloading of catch can be accomplished by using fish pumps. Pumping has an additional advantage in that the fish may be pumped from one RSW tank on board the vessel to another at the shore plant without any additional handling and appreciable increase in temperature. The seawater chilling method ensures that a steady flow of chilled water into every corner of ther fish hold can chill and maintain the fish body uniformly. If the chilling pipe is immersed and fixed in fish hold, the fish body is more vulnerable to discolouring and deterioration because the fish body comes in contact with the chilling pipe (evaporator) more and more as the number of fish increases. Moreover, the farther is the fish body from pipe, the more the fish body becomes chilled. On the contrary the closer the fish body is to the pipe means that the other fish bodies become less chilled. Thus various problems are inherent in this method. The introduction of seawater circulation method can chill fish freshly and it can be changed over for chilling of more than one fish hold by the arrangement of seawater pipe and changing valve between seawater chiller (evaporator) and fish hold (SEAFDEC 2005).

RSW has been found to be of some advantage in improving the quality of fish frozen at sea. On large trawler vessels there are often long delays between catching and freezing. If the fish are left at ambient temperature during those delays, they can undergo rapid spoilage, particularly in warm climates. Because of handling involved when chilling with ice, RSW often offers a more practical solution. In addition, fish that are not bled properly before being frozen often yield pinkish–brownish discoloured flesh, which makes the fish less valuable. Refrigerated sea water allows the fish the time to bleed adequately, while at the same time chilling the fish prior to freezing. A serious problem with RSW and CSW systems is ensuring that tanks, pipes and associated equipment are cleaned thoroughly after each use. Tank holding of fish generally has a high ratio of fish to water, which means that a very high bacterial load may be present when the tanks are loaded with fish. Commercial cleaners and disinfectants are available for cleaning RSW and CSW systems after use.

There are two disadvantages to RSW and CSW systems that affect quality. The first is that in some species the colour of the flesh becomes bleached out when stored for long periods. The second one, and the more serious, is that fish absorb salt from the RSW solution and come become excessively salty if left in too long. This is less prevalent in CSW systems since the brine is diluted somewhat by the melting freshwater ice. Salt penetration may be reduced by lowering the brine to fish ratio. The penetration of salt into fish flesh has discouraged the use of RSW in many fisheries. The amount of salt absorbed depends upon size, species, whether or not the fish have been gutted, the ratio of fish to water, and the length of time in storage (Wheaton & Lawson 1985). Greater weight gain, higher salt levels, higher expressible moisture levels, higher nucleotide levels and faster bacterial growth were observed in salmon held in CSW when compared to iced fish (Crapo et al. 1990).

RSW systems have been used for sardine, salmon, halibut, menhaden, shrimp, mackerel, herring, blue whiting, and many other species. The most successful commercial projects have been confined to bulk applications where the fish are to be used for canning or other industrial processes. Some species, herring for example, keep as well, or a little better, than in ice for 3–4 days, but thereafter spoil more quickly; and some species take up unacceptable amounts of water and salt when kept in sea water. Other species, capelin for example, are reported to keep better in ice, even during the first few days. For these reasons the method is usually confined to short term storage of particular species that are caught in large quantities within a short time, for example herring, mackerel, sprats and blue whiting, since these fish are usually too small and too numerous for gutting or sorting for size to be practicable prior to stowage (Kelman 1977). Salmon could be stored in refrigerated sea water at −0.5°C for 4 days (Crapo & Elliot 1987).

RSW systems can be used with advantage; some of more successful commercial applications are briefly as described:

1 Industrial fishes are chilled in RSW systems to maintain the quality until such time as they are unloaded for processing into fishmeal. Previously, the fish were processed within a day of capture, but longer trips have made it necessary to cool the fish in order to keep them firm and suitable for processing. It is advised that, if a commercial-scale application is contemplated, a prior investigation of all the factors should be made, taking into account seasonal variations in the quality of the fish concerned and the intended end product. The method has been used for storing and transporting large quantities of salmon prior to processing into a canned product. In this application salt uptake is relatively unimportant and the ease of handling, usually by brailing, gives the system an advantage over iced storage. Industrial fish such as menhaden are chilled in RSW systems to maintain the quality until such time as they are unloaded for processing into fish meal. Previously, the fish were processed within a day of capture, but longer trips have made it necessary to cool the fish in order to keep them firm and suitable for processing.

2 Purse seiners. Purse seine fishing vessels use RSW systems for chilling catches, mainly of pelagic fish. Unlike drifters, who bring the catches slowly on board, purse seiners have large catches, which require handling and chilling quickly. The fish are therefore collected from the net directly into RSW tanks.

3 Large-size trawler vessels RSW systems are often used for pre-cooling. RSW systems are often used on factory trawlers when there are likely to be delays between catching and processing. Fish stored in bulk and unrefrigerated between catching and processing will deteriorate quickly, especially in warmer climates (SEAFDEC 2005; Graham et al. 1992).

Clearly, the above applications cover a wide range of circumstances depending on the species of fish and the prevailing climatic conditions; it is difficult to generalize on both the description and use of RSW systems. Salt uptake is probably the most important factor that limits the application of RSW systems. Fish intended for normal processing and marketing can acquire a salt fish taste, which would make them unacceptable. The salt uptake in industrial fish is also critical since it is concentrated during processing. The upper limit is usually equivalent to a concentration of about 0.5% in the raw fish. Salt uptake depends on species, size of fish, salt content of the RSW, ratio of RSW to fish, time, and temperature. Another element that dictates the limit of salt uptake is the preference of the consumer. Therefore, acceptability limits may have to be established not only according to the species and the end product but also in relation to the tolerance of the consumer (Johnston & Nicholson 1992).

4.2.1.4 Chilling with dry-ice

The most important means of food preservation of fresh fish in tropical and temperate regions is by chilling to about 0°C. The most common chilling media is wet ice. However, the quantity of crushed ice required for chilling fresh fish is quite substantial which is at least 1:1, and sometimes is even higher with tropical conditions. The other disadvantages of using water ice are more drip-loss, textual toughness, nutrient loss, and decreased protein extractability Application of dry-ice (solid carbon dioxide) as a novel chilling medium is found to be effective for preserving fish by reducing the temperature rapidly. Dry-ice as a cooling agent has certain advantages, it has bacteriostatic effect, and it acts as an insulant enveloping the fish upon evaporation (Jeyasekaran et al. 2004).

The cryogenic cooling techniques are based on various liquefied gases such as nitrogen (N_2) and helium (He), which are filled in a Dewar container. Even though the liquefied gases such as the liquid phase of N_2 (77 K), Ne (27 K), He (4 K), methane (CH_4), ammonia (NH_3) and carbon dioxide (CO_2) are known as the suitable coolants for various applications; however, among them CO_2 possesses several advantages, including relatively low cost, easy handling, simple transportation and long-term preservation. In fact, CO_2 can be permanently kept in liquid phase using a well-insulated reservoir equipped with an intra-heater and chiller to control the pressure, typically in the interval of 16–20 bar, for four season applications. The other cryogenic liquids, such as N_2 or He, lose mass by evaporation, with a definite loss rate. In general, the handling problems and high cooling costs of liquefied gases restrict their applications to the medical treatment, imaging and the cryogenic research purposes. Carbon dioxide, as a non-polar molecule, possesses a simple structure. It is a colourless gas at ambient conditions, which can be supplied in solid and liquid phases as well. Liquid CO_2 diffuses through a nozzle at atmospheric pressure to turn out into the solid (carbonic snow) with a density reduction of 1.66 kg/m³. Carbonic snow, or dry-ice, is a non-toxic and non-corrosive material, which is attractive for multiple uses where the weight of freight is critical. It is particularly suitable for food transportation and airline catering. In addition, different sizes such as pellet, nugget, slab and block are available for a diversity of demands. The latent heat for the sublimation is 573.1 kJ/kg, which makes it possible to expand up to 800 times of its initial volume. The dry-ice chilling power is notably higher than the ordinary ice of the same mass. Sublimation at atmospheric pressure is the major advantage of dry-ice so as not to wet the surrounding area, leading to a clean and hygienic environment (Ghazaani & Parvin 2011).

Cooling agents such as dry-ice (CO_2 solid) or liquid nitrogen are quick ways of decreasing product temperature in minimum time. Dry-ice sublimates at −78°C and liquid nitrogen evaporates at −195.9°C. Storage of fishery products by using CO_2 solid (CO_2 snow or dry-ice) results in a combined effect of low temperature and CO_2 on the spoilage microflora of the fish. The prospect of this method in transportation of high quality fish and fishery products to distant markets and saving is great because of the longer lasting effects of dry-ice chilling in comparison with other methods (Bao 2004). Because the gas is compacted, then in the process dry-ice will change phase from solid to gas form without passing through the liquid phase, so it can be said that the dry-ice has a cooling effect that is greater than ice cubes made from water (Semin *et al.* 2008). One kilogram of solid carbon dioxide sublimes at atmospheric pressure giving a refrigeration effect 70% larger than 1 kg of ice blocks. Thus, dry-ice has a much greater cooling capacity than the ice blocks of the same size. If the ice cubes capable of absorbing 80 kcal/kg, the dry-ice can absorb 136.6 kcal/kg. Carbon dioxide has long

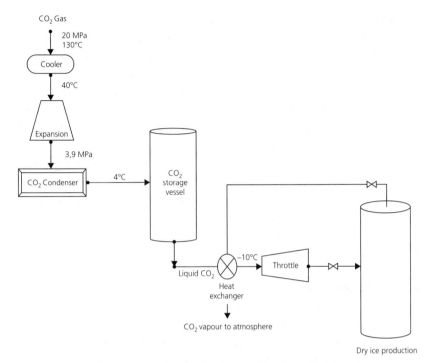

CO$_2$ Gas

20 MPa
130°C

Cooler

40°C

Expansion

3,9 MPa

CO$_2$ Condenser   4°C   CO$_2$ storage vessel

Liquid CO$_2$ Heat exchanger   −10°C   Throttle

CO$_2$ vapour to atmosphere

Dry ice production

Figure 4.3 Dry ice production.

been used for air regulation on meat cold storage; its main function is to prevent bacterial growth. Dry-ice is made with compressed CO$_2$ gas to produce hot, high-pressure gas. Hot gas then cooled to condense into liquid CO$_2$, which is still at high pressure. The liquid is then reduced to 1 atmospheric pressure through spray equipment to produce snow compressed into dry ice crystals are ready for use (Semin & Ismail 2011). Dry ice production system is shown in Figure 4.3.

Dry-ice is very suitable for rapid transportation of fresh fish by air. Recently, exporters use dry-ice combined with wet ice. However, use of dry-ice alone is very expensive, and therefore the combination of dry-ice with water ice is generally recommended. The use of dry-ice alone was found to be cost prohibitive, as a higher concentration was required for effective chilling, while its combination with water ice at lower concentrations was found to be effective, economical, and best suited for the short-term preservation and transportation of high value fishes like seer fish (Sasi *et al.* 2003).

Jeyasekaran *et al.* (2004) stored letrinus (*Lethrinus miniatus*) in a combination of dry-ice and wet ice and they found excellent condition up to 24 h without re-icing. They also stated that there was about 15–25% reduction in the cost of air transport of chilled fish when packed with a combination of dry-ice and water ice. Indian white shrimp preserved with dry-ice and a combination of dry-ice and water ice exhibited an improvement in the quality and shelf life of shrimp by about 6 h when compared to their storage in only water ice (Jeyasekaran *et al.* 2006a). The same researchers found similar results for grouper (*Epinephelus chlorostigma*) (Jeyasekaran *et al.* 2006b). The squid (*Loligo duvaucelli*) tubes stored in 100% dry-ice had a longer shelf life and better quality when compared to their storage in water ice (100%) as well as in the combination package of dry-ice (20%) and water ice (50%) on the basis of important quality indicators (Jeyasekaran *et al.* 2010).

4.2.1.5 Super-chilling

Super-chilling is a concept where the product temperature is reduced 1–2°C below its initial freezing point. During processing, a thin layer of ice is produced on the product surface, and during storage, the ice equalizes within the product and the ice serves as a heat sink. The amount and distribution of ice in super-chilled products prior to further processing greatly affects the processing capacity and yield as well as product quality (Stevik & Claussen 2011; Kaale *et al.* 2013). At super-chilling temperatures, microbial activity is reduced and most bacteria are unable to grow. The rate of food spoilage processes depends on temperature. To reduce spoilage and biochemical degradation, different preservative methods, mainly based on low temperature, have been employed for storage and distribution of food products. The most used methods include refrigerated ice storage between 0°C and 4°C, super-chilled storage in the range of 1 to 4°C, by means of slurry ice or in super-chilled chambers without ice, and frozen storage at −18 to −40°C. The super-chilling/partial-freezing process has two stages:
1 Cooling the product to initial freezing point and
2 Removing the latent heat of crystallization.

The phase transition stage of the super-chilling/partial-freezing process involves the conversion of water to ice through the crystallization process and is the key step in determining the efficiency of the process and the quality of the resulting super-chilled product. The degree of super-chilling (ice fraction) is amount of water (5–30%), which is frozen inside the food product, and it is one of the most important parameters which define the quality of the super chilled food product.

During the storage time, the ice formed will absorb heat from the interior and eventually reach equilibrium. Super-chilling provides the food product with an internal ice reservoir so that no external ice is required during transportation or storage for short periods. Frozen foods have a shelf life of months or years while fresh products have a shelf life of days or weeks. The shelf life of super-chilled food is far shorter than that of frozen food but longer than that of chilled food. The super-chilling technology combines the favourable effect of low temperatures with the conversion of some water into ice, which makes it less available for deteriorative processes. Generally, super-chilling is positioned between freezing and refrigeration (conventional chilling), where the surrounding temperature is set below the initial freezing point. The initial freezing points of most foods are between 0.5°C and 2.8°C (Kaale *et al.* 2011).

Super-chilling, as used for preserving seafood, has been defined as the lowering of the temperature of the flesh to within the range from −3° to −1°C. Pure water freezes at 0°C, but its freezing point is depressed when it contains dissolved substances. The water in biological systems (plants and animals) contains varying amounts of dissolved substances; therefore, the freezing of seafood occurs below the freezing point of pure water. When the temperature of seafood is lowered, the physical change to a hardened mass occurs gradually at rates that are fastest in the beginning and slower as the temperature drops. The water in the seafood is not spontaneously frozen at any given temperature. The first record of super-chilling was reported in about 1935 (Carlson 1969) and it involved the use of brine (at about −3°C) as the refrigerant, resulting in extended shelf lives for whole fish. In the first major use of super-chilling, mechanical refrigeration was used to hold fish aboard fishing vessels at about −1.1°C. Super-chilling is effective and practical, provided that the temperature does not fall below the point where freezing is discernible (about −2°C) and that the time of holding does not exceed 12 days (Ronsivalli & Baker 1981). Several researchers reported that super-chilling extended the shelf life of fish (Sivertsvik *et al.* 2003; Duun & Ruestad 2008; Wang *et al.* 2008; Olafsdottir *et al.* 2006), prawn (Ando *et al.* 2004) and squid (Ando *et al.* 2005).

Super-chilling prior to MAP increased the bacterial inhibition of salmon fillets (Hansen *et al.* 2009). However, the sub-zero temperature zone may influence enzymatic reactions, as substrate concentration increases following partial-freezing of the water phase, which may lead to an altered spoilage process (Lauzon *et al.* 2010). Foegeding *et al.* (1996) stated that it should not be assumed that lowering the temperature to a value close to 0°C necessarily reduces the activity of all enzymes in muscle tissue. Although a reduction in muscle temperature from about 40°C (close to that living mammals) to 0°C lowers the rate of most enzyme-catalysed reactions, it should be realized that some fish live in a cold environment (body temperature equals that of environment) and their enzymes will be more active at low temperatures than will those of warm-blooded animals. Bahuaud *et al.* (2008) reported that super-chilling accelerated the release of these proteolytic enzymes from the lysosomes, causing an acceleration of salmon muscle degradation. These lysosomal breakages seemed to be due to freezing damage, confirmed by the formation of ice-crystals during super-chilling.

The super-chilling process can be carried out in special cold-producing machines called freezers; mechanical freezers, cryogenic freezers or impingement freezers. The three technologies, and all freezers, have different advantages, drawbacks and limitations. The selection of suitable freezing equipment helps to maximize product quality, operating flexibility and return on investment (ROI) while minimizing waste, costs and downtime (Kaale *et al.* 2011). Super-chilling (partial-freezing) at the plant can be done in a freezing tunnel (blast air or cryogenic gases), where the fish holding time in the tunnel is long enough to facilitate super-chilling of the outer layer of the fish or fillet. Alternatively, the fish can be super-chilled using dry-ice, immersed in RSW, or in brine (slurry). For most fish species the freezing point of the muscle varies between 0.8 and 1.4°C (Sikorski 1990). Mechanical freezers are commonly used to freeze foods. Mechanical freezers, especially in continuous belt freezers, have lower operating costs than cryogenic freezers. However, mechanical freezers require higher processing times due to low heat transfer coefficients, which, in turn, lead to a lower quality product. Cryogenic freezing offers shorter freezing times compared to conventional air freezing because of the large temperature differences between the cryogen and the product surface and the high

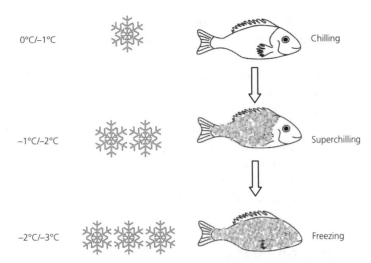

0°C/–1°C   Chilling

–1°C/–2°C   Superchilling

–2°C/–3°C   Freezing

Figure 4.4 Differences of chilling, super-chilling and freezing.

rate of surface heat transfer resulting from the boiling of the cryogen (Kaale *et al.* 2011). Differences of super chilling from chilling and freezing are schematized in Figure 4.4.

The necessary holding time inside the chilling equipment is shortened compared to traditional chilling, as the ice formed under the surface accumulates heat from the interior of the products after leaving the equipment. Assuming there is still some ice content left after equilibrium, the product will have a reservoir of cold to absorb heat from the surroundings, without altering the product temperature too much. Both these advantages are used in industrial applications today. Furthermore, several cutting processes in the food producing industry depend on a partial frozen surface in order to get straight cuts and a stable and reliable production. The amount of heat to be removed increases rapidly as the temperature is lowered below freezing point, due to the rapid formation of ice. During super-chilling, considerably more heat has to be removed compared to normal chilling temperatures can be used due to rapid energy transfer when ice is frozen on the product's surface. The main challenge is to select optimal process conditions, fluid temperature and velocity, and to control the holding time in the super-chilling unit. Quick, low-temperature shell freezing and high heat transfer rates will probably initially result in small ice-crystals. The heat transfer rate may be restricted downwards by a surface temperature that gives excessive freezing of intercellular water, mechanical stress and boosted drip-loss. Therefore, the optimal process and degree of super-chilling depend on the type of food product, its size and geometry, and its intended use or processing after storage (Magnussen *et al.* 2008).

Super-chilling has been regarded as a means to prolong shelf life. Obviously, the sub-zero storage temperature must be taken into account. For example, if the shelf life of a certain fish product in ice is 14 days, the predicted shelf life at 1, 2 and 3°C would be 17, 22 and 29 days, respectively (Huss, 1995).

4.3 Chilling on board

Proper handling and storage of fish at sea ensures that the catch stays as fresh as possible until it is landed. The important requirements are to chill the fish rapidly as soon as they are caught, to keep them chilled, and to maintain a good standard of cleanliness on the fishing deck, in the handling area or shelter deck, and particularly in the fish room or stowage area. A well-designed vessel can make handling of the catch easier. Good stowage practice can help maintain freshness of the catch even in badly designed vessels, or in small boats where stowage facilities are primitive. Bad handling, even on a well-designed ship, can only result in poor quality fish. The importance of good practice at sea cannot be over-emphasized because fish begin to spoil as soon as they die. Neglect on board, even on short fishing trips, can sometimes result in poor-quality fish after only a few hours. Moreover, since the time the fish is on board ship is often longer than the time on shore between landing and consumption, the fisherman may bear much of the responsibility for the freshness of the fish reaching the consumer. In many countries there are now schemes for inspecting and grading the catch on landing. Therefore, the care with which the catch is stowed, as well as the length of the trip, will affect the value put upon the catch. Under these conditions there is usually a financial incentive to the fisherman to bring back the catch in top condition, since the penalty for poor practice may well be downgrading or even withdrawal of the catch from sale. The art of good stowage can vary to some extent with the species being handled, the type of fishery being pursued, the size of vessel and the length of voyage. There are, nevertheless, certain broad principles that apply to almost all fisheries and these are outlined here. Although the advice is based mainly on that given to the north Atlantic trawling industry, most applies equally to the smaller inshore vessel, whether fishing in temperate or tropical waters (Graham *et al.* 1992).

4.3.1 Icing on board

The time between the capture and death of the fish to when they are properly iced must be as short as possible, with minimum exposure to high temperatures. In tropical conditions, this would also require that fish be kept in the shade and out of direct sunlight (Figure 4.5). Without insulation in the fish hold, the rate of ice melting and loss is likely to be high, particularly in tropical and subtropical regions. One way of combating of this without insulating the hold is to place plenty of extra ice against the vessel sides before stowing the fish, and extra layers of ice on top of the last layers of fish near the deckhead, which should help compensate for heat penetration (Shawyer & Pizalli 2003).

The use of seawater ice for chilling fish has been studied for several years and, with the development of suitable small ice machines that

Figure 4.5 Spoilage of fish exposed to sunlight.

can be installed on board fishing vessels; this alternative is becoming more feasible for fishermen. The main advantages of the use of seawater ice are:

• It can be produced at sea or on shore where shortages of freshwater are a serious problem or where freshwater is expensive.
• Since space on fishing vessels is limited, the ability to produce ice when and if it is needed, rather than having to predict needs before a fishing trip begins, can have practical advantages.
• Slightly lower storage temperatures can be obtained with seawater ice; therefore the shelf-life of fish can be prolonged. Commercially available flake/scale ice machines can manufacture seawater ice with a temperature from −9°C to −20°C and a variable percentage of salt content.

However, there are some major disadvantages, such as:

• Seawater ice is not homogeneous, and when stored it can become a mixture of ice-crystals and chilled salt solution, which is semi-fluid in consistency and leaches out the brine solution as the ice rises in temperature. Therefore, seawater ice has no fixed melting point (−1.5°C to −2°C for seawater ice having a salt content between 3 and 3.6%) and losses through melting and leaching of the brine solution will depend on the storage temperature.
• Because of its variable temperature, there is a risk of partially freezing fish and salt absorption (particularly with thin-skinned fish) when using seawater ice.
• Machines specifically designed for seawater ice production are needed to obtain the best-quality ice. These tend to be more expensive to purchase and run than ice machines designed for freshwater ice manufacture. However packaging fish in ice onboard small fishing vessels, whether in boxes shelves, or pounds, is labour-intensive task and other methods have been introduced to reduce the time and labour required (Shawyer & Pizalli 2003).

The following design factors for on-board seawater ice machines should be considered:

• The plant needs to be capable of operating and producing ice under extreme pitching and rolling conditions of fishing vessels.
• The plant needs to be made from non-corrosive materials (such as high quality stainless steel, aluminium, plastics, rubber and fibreglass) to resist the marine environment.

- The equipment needs to operate at a lower temperature than freshwater ice machines, usually between −18°C and −21°C because seawater freezes at a lower temperature than freshwater.

The advantages of having on-board ice machines, especially for fishermen dedicated to the production of fresh fish, can be summarized as follows:

- They allow flexibility in catch volume and trip length.
- After the initial purchase costs of the machine, ice production can be less costly and only involves keeping the ice machine properly maintained and in good repair.
- The fisherman is no longer dependent on shore-based plants for ice supplies for fishing trips; ice can be generated as and when required.
- Being able to produce ice on board can overcome the problems that occur when a boat that has been loaded with shore-side ice returns with little or no catch. Ice costs can amount to a considerable percentage of operational costs in many countries.

The principal disadvantages are:

- Costs of purchase and installation of machine and any ancillary equipment that may also be required, such as auxiliary power, conveyors, etc.
- The ice produced is usually from saltwater, which can affect some fish species by salt absorption into the product.
- Ice, and consequently, the catch can be contaminated if care is not taken to use only clean seawater.
- Machine maintenance will require some specialized technical expertise.
- Additional power is needed.
- Skilled labour and maintenance services are required (possibly on board the vessel) (Shawyer & Pizalli 2003).

When installed on board fishing vessels, flake-ice machines are often mounted on the deck so that the ice produced is discharged directly into the fish hold via a small hatch provided for this purpose. Most drum-type ice-makers designed for fishing vessels have an ice discharge port directly below the drum centre, making installation over a dedicated hatch possible. Depending on the machine, its location on deck and manufacturers' recommendations, some form of shielding or cabinet may be necessary to protect control panels or other parts of the unit from the environment. The below-deck installation is

generally more problematic, as most machines rely on gravity after removal of ice from the drum to put ice in the storage bins. This would require a fairly large fish hold with sufficient height to the deckhead to provide room for the machinery installation and enough height to allow gravity feed to a collection area or storage pens. Flake or shell ice machines may require the installation of conveyors or augers in larger vessels, though in the majority. The heat transfer takes place mainly between the fish and ice and the cold melt water in direct contact. Thus, to achieve the highest chilling rate, it is necessary to surround each fish completely by ice. Although this is possible in laboratory experiments, the commercial conditions on fishing vessels do not always allow for such individual treatment of all fish. Icing on board is traditionally carried out manually. On vessels catching small fish such as herring, mackerel and anchovy, the fisherman have to stow the fish from the individual hauls quickly into the hold to make the deck clear before the next catch. To ensure maximum contact of ice with the fish, proper selection of the size of ice particles and good stowage practices are needed. The rate of chilling is governed by:
• the size, shape and thickness of fish
• the method of stowage
• adequate mixing of ice, water and fish (in ice slurries); adequate contact of ice with the fish
• the size of the ice particles (Shawyer & Pizalli 2003).

4.3.2 Bulk stowage

For bulk stowage of iced fish, the traditional fish room is divided into pounds by fitting removable boards into vertical stanchions. The fish and ice are mixed so that each individual fish is completely surrounded with ice, insuring maximum storage life. The fish are stored in pounds between layers of ice and are separated by ice from the shipside and other fish room structure. Fish in contact with one another will not cool as rapidly as when each fish is completely surrounded by ice. In addition, when fish are stowed so that they are in contact with one another, part of bulkhead, or the side of a box, air may be excluded. Some anaerobic bacteria are capable of producing foul odours, which may rapidly spread throughout the fish flesh. Lean, as well as fatty fish, may suffer from this problem. Such fish are called 'stinkers' or 'bilgy' fish. Fish properly surrounded by ice do not become spoiled in this manner since there are numerous pockets of air trapped between particles of ice. Supporting shelves should be spaced at intervals no greater than 0.5 m. A layer of ice about 5 cm thick should separate the fish from the shelves, top and bottom. A layer of ice should also separate the fish from the sides. The shelves should not be overfilled and care should be taken to make sure that the shelves rest on their supports, not on the bulk of fish and ice immediately blow. This is to avoid crushing.

In bulking, the depth of the bulk should be no greater than about m because the bottom fish may be crushed and suffer excessive weight loss due to loss of fluid. Small fatty fish such as sardine, herring, and related species, do not withstand bulk stowage in ice very well since they are too easily crushed. The main method of stowing of these species is in containers with ice or refrigerated sea water. To avoid heavy losses due to high pressure in the bottom layers of fish, a second platform is formed at about 50 cm from the bottom of fish room, on the outside air temperature and duration of trip. The practical requirements must be found by experience for the individual vessel and type of fishery. For unloading the iced fish baskets of containers are usually used. Much manual labour is required, unless fish pumps or elevators and conveyor systems are employed. The bottom of the fish room should be covered with a layer of ice 10–15 cm deep. The actual depth of the ice depends on how well the fish room is insulated, the length of the voyage, and the temperature outside the fish room. The thickness of the bottom layer should be increased if the floor is made of metal or if it is uninsulated. If this is the case, the bottom layer of fish will have warmed and may be spoiled. Bulked fish are difficult to unload since a large amount of labour is required, and limited mechanization exists. The most common procedure is to separate the fish and ice aboard the vessel, and to load the fish into boxes. The ice is discarded overboard. Another method is to separate the fish and ice on shore. Conveyor belts that automatically separate fish from ice are coming more into use. Pumping has been employed to remove fish and ice from the holds of vessels, but in some cases there has been physical damage to the catch. Some systems are used where fish were mixed with water and pumped out with a centrifugal pump. Pumping systems are widely used with smaller pelagic species (Wheaton & Lawson 1985; Sikorski 1990).

4.3.3 Shelved stowage

In shelving, fish are stored in single layer on a bed of ice. This method of stowage is reserved for the larger species which are gutted and they are placed on the ice with gut cavity down. Sometimes ice is placed on top. Spoilage is retarded since the gut cavity is kept cold. Fish stowed in this manner often have a more attractive appearance than fish bulk stowed. Shelving is laborious and requires double the space required for conventional bulking. The overall quality of fish stowed in this manner may be inferior since icing is incomplete, allowing higher storage temperature (Wheaton & Lawson 1985).

4.3.4 Boxed stowage

Boxing of fish generally results in better quality fish than the other methods of ice stowage. Better control of temperature can be maintained from the time the fish is boxed with ice to the time it is discharged on shore. Often fish remains in boxes until they reach the retail market. This has obvious advantage in terms of quality. If each box is emptied and the contents checked for quality and weight, the catch will be distributed and handled to the point where quality will suffer. Reusable containers used for boxing at sea may be made of plastic-coated wood or plastic. Uncoated wooden boxes are undesirable since it is difficult or impossible to completely disinfect the porous surfaces. Microorganisms present will contaminate the next batch of fish to be stored. With plastic boxes, care should also be taken to minimize the surface area of contact the fish with plastic; however, plastic boxes are preferred where fish rooms are uninsulated. Non-returnable boxes may be made of wire-bound wood, nailed wood, wax-impregnated corrugated fibreboard or foamed polystyrene. Containers used for boxing are generally about 70 kg in capacity so that they may hold both fish and ice. Containers of this size simplify the problem or grading fish at the market for size, species or freshness. Fish of similar size and species can be placed into boxes aboard the vessel and grouped into lots for sale once discharged from the vessel. The freshness of all of the specimens in one box should be the same. Another advantage is that removing the boxed fish from the vessel is much easier and less laborious. Design of the box is important. One box should be large enough to hold sufficient quantities of fish and enough ice to chill the catch properly and keep it chilled until landed. A sufficient large box should be selected. A common fault is to use insufficient ice because the boxes are not large enough. The depth of box should be sufficient so that bottom fish are not crushed. Drain holders should be placed on the ends or sides of the boxes so that melt water does not drain down through the fish below (Wheaton & Lawson 1985).

4.3.5 Quantity of ice needed on board

The amount of storage space required for ice will depend on the size of vessel, the length of fishing, the quantity of fish lightly to be caught and whether the vessel has onboard ice making facilities. Careful consideration must be given to how much ice is actually necessary for the average trip. Several factors should be considered when trying to estimate adequate quantity of ice per trip such as

- planned length of trip
- historic average catches per trip
- type of fish being caught (large medium, small)
- available carrying space in hold and/or containers
- anticipated ice losses to heat gain in hold or containers
- local ambient temperature.

The market for fish being caught may also have a bearing on the quantities of ice to be used. Ideally, there should still be some ice left in the hold after all fish have been discharged at the end of each trip. This will indicate that the weight of ice was correctly estimated. The amount of ice that can be taken on board may be restricted by the space availability for initial storage before commencement of fishing. A balance has to be struck between ice carried, anticipated catch, and catch composition. In practice, the carried could be expected to exceed the minimum estimated quantities by 30% or more, to compensate for heat losses.

4.3.6 Use of refrigerated seawater

RSW systems have an on-board refrigeration plant to chill the seawater rather than using melting ice. In addition, they need pumps, piping and filter for circulation of the RSW in the tanks or holds. In normal practice this system requires a dedicated power plant, such as a diesel or diesel electric generator, providing direct power or electricity to operate the electric motors for refrigeration compressors and circulation pumps, depending on the type of drive motors used. Two basic systems are used for RSW cooling of products: one involves simply immersing the catch in filled RSW tanks; the second system does not use tanks but sprays chilled water over shelved catch. When filling RSW tanks in the hold with clean water that is then refrigerated, some boats will load ice into the tanks prior to filling with water. This saves time and alleviates some of the load on the refrigeration system by pre-chilling the water. (Shawyer & Pizalli 2003).

4.3.7 Use of chilled seawater

CSW is very efficient in cooling because of the fish are completely surrounded by the cooling medium. However, it requires watertight boxes or tanks installed in the hold that can incur extra costs which cannot always be justified. This type of installation is mostly used for:

- Fisheries dedicated to capture of high value species
- The preservation of fish where there is a relatively short time between capture and delivery to processing plants
- The preservation of bulk catches of small pelagics where it is impractical to ice fish individually. In the case, fish can often be loaded directly from purse seine into CWS hold thus providing rapid and efficient chilling (Shawyer & Pizalli 2003).

4.4 Combination of chilling with traditional and advanced preserving technologies

Whatever the processing of fish receives the temperature will tend to rise. This can be kept in check in between certain stages, such as gutting and filleting, by icing, and by immersion in chilled water. However, in the production of drier products such methods are not suitable and the fish are best cooled in chill stores that use cold air circulation to cool the product. In dispatching the end products two methods of chilling are used. The first relies on packaging with ice. With wet products this is an efficient method and, when used in conjunction with an insulated container and sufficient ice, it can keep the fish at 0°C for 2–3 days depending on the outside environment. Dried products such as smoked fish cannot be chilled in this manner. They must be cooled and stored in refrigerated chambers (Whittle et al. 1990).

Chilling extends the shelf life of seafood preserved or processed by other techniques. Fresh fish products have a limited shelf life, which is an important hurdle for the export of fresh products. Handling, cooling, processing, packaging and storage methods greatly influence the quality maintenance and successive deterioration process of such products. Raw material quality and other intrinsic/extrinsic parameters related to the fish influence the freshness and shelf life extension that can possibly be achieved. Temperature control is important to maintain fish quality. Pre-cooling of fillets in process is being used to lower the temperature prior to packaging (Lauzon *et al.* 2010). Chilled storage is based on lowering temperature close to the freezing point of the products and is a useful technique that has been applied to extend the shelf life of fish and fish products. The low temperature will reduce the activities of spoilage bacteria, enzyme activity and lipid oxidation reaction, so the fish and fish products remain edible longer. Multiple interactions between biological, physical and technical factors affect the efficiency of chilled storage.

Modified atmosphere packaging (MAP) can be combined with super-chilling to further extend the shelf-life and safety of fresh fish. In this technique, also known as partial- freezing, the temperature of the fish is reduced to 1–2°C below the initial freezing point and some ice is formed inside the product and can be used to extend prime quality of fresh fish (Siverstvik *et al.* 2002). Synergism of combined super-chilling and MAP can lead to a considerable extension of the freshness period and shelf life of fish products. Low storage temperature and modified atmosphere packaging (MAP) inhibit bacterial growth and biochemical degradation. Super-chilling prior to MAP increased the bacterial inhibition of fish fillets (Hansen *et al.* 2009). Hurdle technology means the intentional combination of preservation techniques in order to establish a series of preservative factors that any microorganisms present should not be able to overcome. These hurdles may be storage temperature, water activity, pH, redox potential preservatives, and novel techniques including MAP. Smart packaging such as time temperature indicators is a technology that appears to have a potential, especially with chill stored MAP products. To ensure microbial safety strict temperature control is needed and temperature abuse should be avoided (Siverstvik *et al.* 2002).

Fish salting and chilling can take place together. The process may be called warm salting, chilled salting or cold salting depending on the amount of ice used. In chilled salting fish is salted after being chilled to 0–5°C. Thanks to chilling, autolytic and bacterial activities on fish tissue are decelerated and fish quality is preserved during storage. In this method, salt and ice are used together in open containers. Fish are mixed with coarse-grained salt and fragmented ice. Penetration of salt into fish gets easier as the ambient temperature increases as a rise in temperature increases the rate of autolysis and the level of bacterial activity. Temperatures below 10°C are appropriate for salting. The best salting is performed between 3° and 4°C. Chilling is very important for storage of salted fish (Turan & Erkoyuncu 2012).

Some antimicrobials and sanitizers can be incorporated in the ice to extend the shelf life of seafood. Good results were obtained for quality and shelf life of seafood stored in ice containing plant extracts such as thyme (Oral *et al.* 2008; Bensid *et al.* 2014), oregano and rosemary (Quitral *et al.* 2009; Bensid *et al.* 2014), clove (Bensid *et al.* 2014). Addition of organic acids to ice and the use of ice with organic acid have been studied. Rey *et al.* (2012) reported that use of ice containing ascorbic acid, citric acid and lactic acid protected microbial and sensory quality and extended the shelf life of hake (*Merluccius merluccius*), megrim (*Lepidorhombus whiffiagonis*) and angler (*Lophius piscatorius*). Flake-ice containing citric and lactic acids also provided inhibitory effect on microbiological, chemical and sensory parameters of hake (Garcia-Soto *et al.* 2011, 2013). The sanitized ice improved the sensory, microbiological and biochemical quality of sardines (Campos *et al.* 2005). *E. coli* O157:H7, *Salmonella typhimurium*, and *Listeria monocytogenes* in fish skin were reduced by using ice containing chlorine dioxide (ClO$_2$) (Shin *et al.* 2004).

Dipping of preservative solution or a coating agent can be combined with chilling. This kind application protects the quality and increase the shelf of fish. Sardine fillets were coated with chitosan and stored with flake-ice. Edible coating with chitosan was effective in inhibiting bacterial growth and reduced the formation of volatile bases and oxidation products significantly. A shelf life of 8–10 days was found for coated sardine fillets. Untreated samples (only flake-ice) had a shelf life of 5 days (Mohan *et al.* 2012). Combination of sodium acetate dip treatment, vacuum/MAP, icing and refrigerated storage at 0–2°C prolonged the shelf life of pearlspot (*Etroplus suratensis*) (Manju *et al.* 2007) and seer fish (*Scomberomorrus commerson*) (Yesudhasan *et al.* 2014). Quality loss and melanosis in pacific white shrimp were retarded by dipping into tea extracts and then storing in ice (Nirmal & Benjakul 2011).

Tea polyphenol dip treatment retained good quality characteristics for longer and extended the shelf life of silver carp (*Hypophthalmicthys molitrix*) during iced storage (Fan *et al.* 2008). Ozone treatment increased effectiveness of icing in Pacific white shrimp (*Litopenaeus vannamei*). Good quality was observed in ozone-treated samples during iced storage compared to alone icing (Odilichukwu & Okpala 2014).

References

Ando, M., Nakamura, H., Harada, R. & Yamane, A. (2004). Effect of super chilling storage on maintenance of freshness of kuruma prawn. *Journal of Food Science and Technology Research*, **10** (1): 25–31.

Ando, M., Takenaga, E., Hamase, S. & Yamane, A. (2005). Effect of super-chilling storage on maintenance of quality and freshness of swordtip squid *Loligo edulis*. *Food Science and Technology Research*, **11** (3): 355–361.

Aubourg, S.P., Losada, V., Prado, M., Miranda, J.M. & Velazquez, J.B. (2007). Improvement of the commercial quality of chilled Norway lobster (*Nephrops norvegicus*) stored in slurry ice: Effects of a preliminary treatment with an antimelanosic agent on enzymatic browning. *Food Chemistry*, **103**: 741–748.

Bao, H.N.D. (2004). Effects of dry ice and superchilling on the quality and shelf life of arctic charr *(Salvelinus alpinus)* fillets. Final Project. Pp 68, UNU-Fisheries Training Prgramme, Reykjavik, Iceland.

Bahuaud, D., Mørkøre, T., Langsrudc, Ø., *et al.* (2008). Effects of 1.5°C super-chilling on quality of Atlantic salmon (*Salmo salar*) pre-rigor fillets: cathepsin activity, muscle histology, texture and liquid leakage. *Food Chemistry*, **111**: 329–339.

Bensid, A., Ucar, Y., Bendeddouche, B. & Ozogul, F. (2014). Effect of the icing with thyme, oregano and clove extracts on quality parameters of gutted and beheaded anchovy (*Engraulis encrasicholus*) during chilled storage. *Food Chemistry*, **145**: 681–686.

Blanc, M. (1995). Fish Handling and Chilling. FAD Fishing Skills Workshop-Module 3.39 p.

Campos, C.A., Rodríguez, O., Losadac, V., Aubourgc, S.P. & Barros-Velázquez, J. (2005). Effects of storage in ozonised slurry ice on the sensory and microbial quality of sardine (*Sardina pilchardus*). *International Journal of Food Microbiology*, **103**: 121–130.

Carlson. C. J. (1969). Superchilling fish – a review. In: *Freezing and Irradiation of Fish* (Ed. Kreuzer, R.), pp. 101–103, Fishing News (Books) Ltd., London.

Crapo, C. & Elliot, E. (1987). Salmon quality: the effects of elevated refrigerated sea water chilling temperatures. *Marine Advisory Bulletin*, 34. Alaska Sea Grant College Program, Fairbanks, AK.

Crapo, C., Himelbloom, B., Brown, E., Babbitt, J. & Reppond, K. (1990). Salmon Quality. In: The effects of ice and chilled seawater storage. Alaska Sea Grant College Program. University of Alaska Fairbanks. *Marine Advisory Bulletin* (MAB) No 40, p. 15.

Duun, A.S. & Rustad, T. (2008). Quality of superchilled vacuum packed Atlantic salmon (*Salmo salar*) fillets stored at 1.4 and 3.6°C. *Food Chemistry*, **106** (1): 122–131.

Egolf, P.W. & Kauffeld, M. (2005). From physical properties of ice slurries to industrial ice slurry applications. *International Journal of Refrigeration*, **28**: 4–12.

Emokpae, A.O. (1982). Icing of fish and construction of an iced fish box. Technical Paper No.1. Nigerian Institute for Oceanography and Marine Research, Victoria, Iceland.

Erikson, U., Misimi, E. & Gallart-Jornet, L. (2011). Superchilling of rested Atlantic salmon: Different chilling strategies and effects on fish and fillet quality. *Food Chemistry*, **127**: 1427–1437.

Fan, W., Chi, Y. & Zhang, S. (2008). The use of a tea polyphenol dip to extend the shelf life of silver carp (*Hypophthalmicthys molitrix*) during storage in ice. *Food Chemistry*, **108**: 148–153.

FAO. (1984). Planning and engineering data 4. *Containers for Fish Handling* (Eds Brox, J., Kristiansen, M., Myrseth, A. & Aasheim, P.W.). Fisheries Circular No. 773, Food and Agriculture Organization, Rome.

Foegeding, E.A., Lanier, T.C. & Hultin, H. (1996). Characteristics of edible muscle tissues. In: *Food Chemistry* (Ed. Fennema, O.R.), 3rd edn, pp. 880–938, Marcel Dekker, Inc., New York.

Garci′a-Soto, B., Sanjua′s, M., Barros-Velázquez, J. Fuertes-Gamundi, J.R. & Aubourg, S.P. (2011). Preservative effect of an organic acid-icing system on chilled fish lipids. *European Journal of Lipid Science and Technology*, **113**: 487–496.

García-Soto, B., Aubourg, S.P., Calo-Mata, P. & Barros-Velázquez, J. (2013): Extension of the shelf life of chilled hake (*Merluccius merluccius*) by a novel icing medium containing natural organic acids. *Food Control*, **34**: 356–363.

Ghazaani, M.I. & Parvin, P. (2011). Characterization of a dry-ice heat exchanger. *International Journal of Refrigeration*, **34**: 1085–1097.

Graham, J., Johnston, W.A. & Nicholson, F.J. (1992). *Ice in Fisheries*. FAO Fisheries Technical Paper No 331. FAO, Rome, Italy.

Hagg, C. (2005). Ice slurry as secondary fluid in refrigeration systems. Fundamentals and applications in supermarkets. Licentiate Thesis. School of Industrial Engineering and Managemet, Department of Energy Technology, Division of Applied Thermodynamics and Refrigeration, Stockholm, Sweden.

Hansen, A.A., Mørkøre, T., Rudi, K., Langsruda, Ø. & Eiea, T. (2009). The combined effect of superchilling and modified atmosphere packaging using CO_2 emitter on quality during chilled storage of pre-rigor salmon fillets (*Salmo salar*). *Journal of the Science of Food and Agriculture*, **89** (10): 1625–1633.

Heldman, D.R. & Hartel, R.W. (1997). Refrigerated storage. In: *Principles of Food Processing*, pp. 88–90. Chapman & Hall, New York.

Huidobro, A., Mendes, R. & Nunes, M. (2001). Slaughtering of gilthead seabream (*Sparus aurata*) in liquid ice : influence of fish quality. *Zeitschriftfür Lebensmittel-Untersuchungund-Forschung*, **213**: 267–272.

Huidobro, A. & Lopez-Cbellero, M.E. (2002). Onboard processing of deepwater pink shrimp (Parapenaeuslongirostris) withliquidice: Effect on quality. *European Food Research and Technology*, **214**: 469–475.

Huss, H.H. (1995). *Quality and quality changes in fresh fish*. FAO Fisheries Technical Paper No 348, FAO, Rome, Italy.

Jeyasekaran, G., Ganesan, P., Shakila, R.J., Maheswari, K. & Sukumar, D. (2004). Dry-ice as a novel chilling medium along with water ice for short-term preservation of fish Emperor breams, lethrinus (*Lethrinus miniatus*). *Innovative Food Science and Emerging Technologies*, **5**: 485- 493.

Jeyasekaran, G., Ganesan, P., Anandaraj, R., Jeya Shakila, R. & Sukumar, D. (2006a). Quantitative and qualitative studies on the bacteriological quality of Indian white shrimp (*Penaeus indicus*) stored in dry-ice. *Journal of Food Microbiology*, **23**: 526–533.

Jeyasekaran, G., Ganesan, P., Anandaraj, R., Jeya Shakila, R. & Sukumar, D. (2006b). Microbiological and biochemical quality of grouper (*Epinephelus chlorostigma*) stored in dry ice and water ice. *International Journal of Food Science and Technology*, **43**: 145–153.

Jeyasekaran, G., Jeya Shakila, R., Sukumar, D., Ganesan, P. & Anandaraj, R. (2010). Quality changes in squid (*Loligo duvaucelli*) tubes chilled with dry ice and water ice. *Journal of Food Science and Technology*, **47** (4): 401–407.

Kaale, L.D., Eikevik, T.M., Rustad, T. & Kolsaker, K. (2011). Superchilling of food, a review. *Journal of Food Engineering*, **107**: 141–146.

Kaale, L.D., Eikevik, T.M., Bardal, T., Kjorsvik, E. & Nordtvedt, T.S. (2013). The effect of cooling rates on the ice-crystal growth in air-packed salmon fillets during superchilling and superchilled storage. *International Journal of Refrigeration*, **36**: 110–119.

Karel, M. & Lund, D.B. (Eds) (2003). Storage at chilling temperatures. In: *Physical Principles of Food Preservation*, 2nd edn, pp. 226–262, Marcel Dekker, Inc., New York.

Kauffeld, M., Kawaji, M. & Egolf, P.W. (2005). Fish storage. In: *Handbook on Ice Slurries- Fundamentals and Engineering*. International Institute of Refrigeration Publication, pp. 3–6, Paris, France.

Kauffeld, M., Wang, M.J., Goldstein, V., K. & Kasza, K.E. (2010). Ice Slurry Applications. *International Journal of Refrigeration*, **33** (8): 1491–1505.

Kelman, J.H. (1977). Stowage of fish in chilled sea water. Torry Advisory Note No 73. Torry Advisory Station, Aberdeen, Scotland.

Lauzon, H.L., Margeirsson, B., Sveinsdóttir, K., Guðjónsdóttir, M., Karlsdóttir, M.G. & Martinsdóttilauzonr, E. (2010). Overview on fish quality research – impact of fish handling, processing, storage and logistic on fish quality deterioration. Technical Report 39–10, Matis, Reykjavik, Iceland.

Light, N. & Walker, A. (1990). An overview of cook-chill catering. In: *Cook-Chill Catering Technolgy and Management* (Ed. Light, N.), pp. 11–17, Elsevier Science Publishers, Ltd, Harlow.

Losada, V., Barros-Velazquez, J., Gallardo, J. & Aubourg, S. (2004a). Effect of advanced chilling methods on lipid damage during sardine (*Sardina pilchardus*) storage. *European Journal of Lipid Science and Technology*, **106**: 844–850.

Losada, V. Piñeiro, C., Barros-Velázquez, J. & Aubourg, S. (2004b). Effect of slurry ice on chemical changes related to quality loss during european hake (*Merluccius merluccius*) chilled storage. *European Food Research and Technology*, **219**: 27–31.

Losada, V. Piñeiro, C., Barros-Velázquez, J. & Aubourg, S. (2005). Inhibition of chemical changes related to freshness loss during storage of horse mackerel (*Trachurus trachurus*) in slurry ice. *Food Chemistry*, **93**: 619–625.

Losada, V. Barros-Velázquez, J. & Aubourg, S. (2007). Rancidity development in frozen pelagic fish: Influence of slurry ice as preliminary chilling treatment. *LWT- Food Science and Technology*, **40**: 991–999.

Magnussen, O.M., Haugland, A., Hemmingsen, A.K.T., Johansen, S. & Nordtvedt, T.S. (2008). Advances in superchilling of food and process characteristics and product quality. *Trends in Food Science & Technology*, **19**: 418–424.

Manju, S., Jose, L., Gopal, T.K.S., Ravishankar, C.N. & Lalitha, K.V. (2007). Effects of sodium acetate dip treatment and vacuum-packaging on chemical, microbiological, textural and sensory changes of Pearlspot (*Etroplus suratensis*) during chill storage. *Food Chemistry*, **102**: 27–35.

Merritt, J.H. (1969). Mechanical refrigeration with ice. In: *Refrigeration of Fishing Vessels*, pp. 57–77. Fishing News Books Ltd., London, UK.

Mohammed, I.M.A. & Hamid, S.H.A. (2011). Effect of chilling on microbial load of two fish species (*Oreochromis niloticus* and *Clarias lazera*). *American Journal of Food and Nutrition*, **1**(3): 109–113.

Mohan, C.O., Ravishankar, C.N., Lalitha, K.V. & Gopal, T.K.S. (2012). Effect of chitosan edible coating on the quality of double filleted Indian oil sardine (*Sardinella longiceps*) during chilled storage. *Food Hydrocolloids*, **26**: 167–174.

Mugica, B., Vela´zquezb, J.B, Mirandab, J.M. & Auborg, S.P. (2008). Evaluation of a slurry ice system for the commercialization of ray (*Raja clavata*): effects on spoilage mechanisms directly affecting quality loss and shelf-life. *LWT-Food Science and Technology*, **41**: 974–981.

Nirmal, P. & Benkajul, S. (2011). Use of tea extract for inhibition of polyphenoloxidase and retardation of quality loss of Pacific white shrimp during iced storage. *LWT-Food Science and Technology*, **44**: 924–932.

Oral, N., Gulmez, M., Vatansever, L., & Guven, A. (2008). Application of antimicrobial ice for extending shelf life of fish. *Journal of Food Protection*, **71**: 218–222.

Pineiro, C., Velazquezb, J.B. & Aubourg, S.P. (2004). Effects of newer slurry ice systems on the quality of aquatic food products: a comparative review versus flake-ice chilling methods. *Trends in Food Science & Technology*, **15**: 575–582.

Price, R.J., Melvin, E.F. & Bell, J.V. (1991). Postmortem changes in chilled round, bled and dressed albacore. *Journal of Food Science*, **56** (2): 318–321.

Odilichukwu, C. & Okpala, R. (2014). Investigation of quality attributes of ice-stored Pacific white shrimp (*Litopenaeus vannamei*) as affected by sequential minimal ozone treatment. *LWT-Food Science and Technology*, **57**: 538–547.

Olafsdottir, G., Lauzon, H.L., Martinsdóttir, E., Oehlenschläger, J. & Kristbergsson, K. (2006). Evaluation of shelf life of superchilled cod (*Gadus morhua*) fillets and the influence of temperature fluctuations during storage on microbial and chemical quality indicators. *Journal of Food Science*, **71** (2): 97–109.

Quitral, V., Donoso, M.L., Ortiz, J., Herrera, M.V., Araya,H. & Aubourg, S. (2009). Chemical changes during the chilled storage of Chilean jack mackerel (*Trachurus murphyi*): Effect of a plant-extract icing system. *LWT – Journal of Food Science*, **42**: 1450–1454.

Rey, M.S., García-Soto, B., Fuertes-Gamundi, J.R., Aubourg, S. & Barros-Velázquez, J. (2012). Effect of a natural organic acid-icing system on the microbiological quality of commercially relevant chilled fish species. *LWT – Food Science and Technology*, **46**: 217–233.

Rodriguez, O., Vanesa Losada, V., Aubourg, S.P. & Barros-Velazquez, J.B. (2004). Enhanced shelf-life of chilled European hake (*Merluccius merluccius*) stored in slurry ice as determined by sensory analysis and assessment of microbiological activity. *Food Research International*, **37**: 749–757.

Rodrıguez, O., Velazquez, J.B., Pineiro, C., Gallardo, J.M. & Aubourg, S.P. (2006). Effects of storage in slurry ice on the microbial, chemical and sensory quality and on the shelf life of farmed turbot (*Psetta maxima*). *Food Chemistry*, **95**: 270–278.

Ronsivalli, L.J. & Baker, D.W. (1981). Low temperature preservation of seafoods: a review. *Marine Fisheries Review*, **43** (4): 1–15.

Sasi, M., Jeyasekaran, G., Shanmugam, S.A. & Jeya Shakila, R. (2003). Evaluation of the quality of seer fish (*Scomberomorus commersonii*) stored in dry ice (solid carbon dioxide). *Journal of Aquatic Food Product and Technology*, **12** (2): 61–72.

SEAFDEC (2005). Chilling systems by refrigerated sea water. In: *Onboard Fish Handling and Preservation Technology*, pp. 6–9, Southeast Asian Fisheries Development Center, Samut-Prakan, Thailand.

Semin, A.R. Ismail, R.A. Bakar & I. Ali (2008). Heat transfer investigation of intake port engine based on steady-state and transient simulation. *American Journal of Applied Science*, **5**: 1572–1579.

Semin, A.R. & Ismail, R.A. (2011). Effect of dry ice application in fish hold of fishing boat on the fish quality and fisherman income. *American Journal of Applied Science*, **8** (12): 1263–1267.

Shawyer, M. & Pizalli, A.F.M. (2003).The use of ice on small fishing vessels. FAO Fisheries Technical Paper No 436. FAO, Rome, Italy.

Shin, J.H., S. Chang, & D.H. Kang. (2004). Application of antimicrobial ice for reduction of foodborne pathogens (*Escherichia coli* O157:H7, *Salmonella typhimurium, Listeria monocytogenes*) on the surface of fish. *Journal of Applied Microbiology*, **97**: 916–922.

Sikorski, E.Z. (1990). Chilling of fresh fish. In: *Seafood: Resources, Nutritional Composition, and Preservation* (Ed: Sikorski, Z.E.), pp. 55–75. CRC Press Inc., Boca Raton.

Sivertsvik, M., Jeksrud, W.K. & Rosnes, J.T. (2002). A review of modified atmosphere packaging of fish and fishery products - significance of microbial growth, activities and safety. *International Journal of Food Science and Technology*, **37**: 107–127.

Sivertsvik, M., Rosnes, J.T. & Kleiberg, G.H. (2003). Effect of modified atmosphere packaging and superchilled storage on the microbial and sensory quality of Atlantic Salmon (*Salmo salar*) fillets. *Journal of Food Science*, **68** (4): 1467–1472.

Stevik, A.M. & Claussen, I.C. (2011). Industrial superchilling, a practical approach. *Procedia – Food Science*, **1**: 1265–1271.

Tuckey, N.P.L., Forgan, L.G. & Jerrett, A.R. (2012). Fillet colour correlates with biochemical status in Australasian snapper (*Pagrus auratus*) during storage in refrigerated seawater. *Aquaculture*, **356–357**: 256–263.

Turan, H. & Erkoyuncu, H. (2012). Salting technology in fish processing. In: *Progress in Food Preservation* (Eds Bhat, R., Alias, A.K. & Paliyath, G.), pp. 297–315, Wiley-Blackwell, Oxford.

Wang, T., Sveinsdóttir, K., Magnússon, H. & Martinsdóttir, E. (2008). Combined application of modified atmosphere packaging (MAP) and superchilling storage to extend the shelf life of fresh cod (*Gadus morhua*) loins. *Journal of Food Science*, **73** (1): 11–19.

Warren, P. (1986). The chilled fish chain. An Open Learning Module for The Sea fish Open Technical Project. Her Majesty's Stationery Office, London.

Wheaton, F.W. & Lawson, T.B. (1985). Quality changes in aquatic food products. In: *Processing Aquatic Food Products* (Eds Wheaton, F.W. & Lawson, T.B.), pp. 239–250, John Wiley & Sons Inc., New York.

Whittle, K.J., Hardy, R. & Hobbs, G. (1990). Chilled fish and fishery products. In: *Chilled Foods* (Ed. Gormley, T.R.), pp. 87–116, Elsevier Science Publishers, London.

Yesudhasan, P., Lalitha, K.V., Srinivasa Gopal, T.K. & Ravishankar, C.N. (2014). Retention of shelf life and microbial quality of seer fish stored in modified atmosphere packaging and sodium acetate pre-treatment. *Food Packaging and Shelf Life*, **1**: 123–130.

CHAPTER 5

Quality changes of fish during chilling

5.1 Introduction

Spoilage of fish is considered as any change that renders the product unacceptable for human consumption. Essentially, fish and seafood products spoilage is a consequence of various microbial, biochemical and chemical breakdown processes. The degree of processing and preservation, together with storage temperature, will decide whether the fish undergoes microbial spoilage, biochemical spoilage or a combination of both. The initial quality loss is mainly due to the post mortem autolytic activity and chemical degradation processes, such as lipid oxidation. However, the most prevalent form in fresh fish chilled or not, is microbial spoilage (Odoli 2009). Information about handling, processing and storage techniques, including time/temperature histories, which can affect the freshness and quality of the products, is very important for the partners in the chain.

Additionally, seasonal condition, the effects of fishing grounds and catching methods and the occurrence of various quality defects influence the overall quality. Fresh fish is susceptible to rapid spoilage at an ambient temperature. A reduction of storage temperature by 5–6°C halves the rate of biochemical reactions and doubles the shelf-life of

Seafood Chilling, Refrigeration and Freezing: Science and Technology, First Edition.
Nalan Gökoğlu and Pınar Yerlikaya.
© 2015 John Wiley & Sons, Ltd. Published 2015 by John Wiley & Sons, Ltd.

fish. The preservation of fish in ice is one of the most efficient ways of retarding spoilage. Proper icing keeps the fish in an acceptable condition for reasonable periods (Jeyasekaran *et al.* 2005). There are many variations in the shelf-life of fish preserved in ice depending on the species, method of capture, location of fishing grounds and size of fish (Lima dos Santos *et al.* 1981). A few hours delay in icing considerably reduces the storage life of fresh fish.

Chilling of fish can slow down the spoilage process, but it cannot stop it. Therefore, it is a race against time and fish should be moved as quickly as possible. The main question for fishermen, traders and consumers is how long fish will keep in ice. Shelf-life will depend on several factors. However, the fish spoilage pattern is similar for all species, with four phases of spoilage. (Shawyer & Pizalli 2003).Of all the factors affecting shelf-life, most interest has focused on the possible difference in iced shelf-life between fish caught in warm, tropical waters and fish caught in cold, temperate waters. Some tropical fish can be kept 20–30 days when stored in ice (Huss 1995). This can be attributed to differences in the bacterial growth rates, with a 1–2 week slow growth phase (or period of adaptation to chilled temperatures) in tropical fish stored in ice (Shawyer & Pizalli 2003). This is far longer than for most temperate species, and several studies have been conducted assessing the shelf-life of tropical species. Several authors have concluded that fish taken from warm waters keep better than fish from temperate waters whereas Lima dos Santos (1981) concluded that also some temperate water fish species keep extremely well and that the longer shelf lives in general are found in fresh water fish species compared to marine species. However, he also noted that the shelf-life of more than 3 weeks, which is often observed for fish caught in tropical waters, never occurs when fish from temperate waters are stored in ice. The iced shelf-life of marine fish from temperate waters varies from 2 to 21 days which does not differ significantly from the shelf-life of temperate freshwater fish ranging from 9 to 20 days. Contrary to this, fish caught in tropical marine waters keep for 12–35 days when stored in ice and tropical freshwater fish from 6 to 40 days. Although very wide variations occur, tropical fish species often have prolonged shelf lives when stored in ice. When comparisons are made, data on fatty fish like herring and mackerel should probably be omitted as spoilage is mainly due to oxidation (Huss 1995).

5.2 Chemical changes

During chilled fish storage a number of inter-related systems take place. Among those factors are changes in protein and lipid fraction, degradation of nucleotides with the subsequent formation of amines (volatile and biogenic), hypoxanthine (Hx) and action of certain bacteria. One of the most important chemical spoilage processes are changes taking place in the lipid fraction of the fish. After death, the lipids in fish are subjected to two major changes, lipolysis and auto-oxidation (Hardy 1980). In chilled fresh fish and even in fish stored at ambient temperatures, lipid oxidation does not seem to be a dominant spoilage process, though in the later stages of spoilage of fattier species such as trout, sardine, herring and mackerel, rancid flavours affect acceptability (Sikorski & Kolakowski 2000; Whittle *et al.* 1990).

The non-protein nitrogen (NPN) (such as free amino acids, total volatile basic nitrogen (TVB-N – i.e. ammonia, trimethylamine (TMA), creatine, taurine, uric acid, carnosine and histamine) content of fish easily causes spoilage. Volatile bases result from degradation of proteins and non-protein nitrogenous compounds. The characteristic odour and the progression of odour in fish during storage has been associated with varying levels of different volatile compounds present in the headspace of fish, which can be measured to evaluate the freshness of fish. The composition of volatile compounds in fish contributing to the characteristic odours can be determined and related to the quality. The value of 30–35 mg TVB-N/100 g is recommended for fresh fish acceptability (Huss 1988; Connell 1995). One of the volatile compounds can be a TMA, which appears by bacterial decomposition of trimethylamine oxide (TMAO). TMAO in fish muscle can be degraded to TMA by endogenous enzymes, but at chill storage temperatures TMA is produced by the bacterial enzyme TMA-oxidase (Ashie *et al.* 1996). When the oxygen level is depleted, many of the spoilage bacteria utilize TMAO as a terminal hydrogen acceptor, thus allowing them to grow under anoxic conditions. Towards the end of shelf-life various odorous low-molecular-weight sulphur compounds such as H_2S and CH_3SH, together with volatile fatty acids and ammonia, are produced because of bacterial growth (Sivertsvik *et al.* 2002). Fish have a unique osmoregulatory mechanism to avoid dehydration in marine environments and waterlogging of tissue in fresh water.

One important osmoregulant is TMAO, found in concentrations up to 1% in teleost and up to 1.5% in elasmobranchs, together with 2% urea. However, Gram-negative bacteria such as *Shewanella putrifaciens* can obtain energy from TMAO by reducing it to TMA. Endogenous enzymes present in fish can also reduce TMAO to DMA (dimethylamine) and FA (formaldehyde). While TMAO is non-odorous, TMA is a component in the odour of stale fish. The levels of TMA in fish has for sometimes been used as an indicator of microbial deterioration of fish (Fraser & Sumar 1998). TMA gives a very characteristic 'fishlike' smell. At the beginning of this decomposition phase, the smell and taste can be mildly sour, similar to the smell of cabbage, ammonia, or unpleasant smells of rotten and rancid fish (Bojanic *et al.* 2009). TMAO is found in marine fishes; the TMA-N content of blue whiting reached to 13.6 mg/100 g after 11 days of storage in ice (Dagbjartsson 1975). While TMA-N values of redfish reached 36 mg/100 g after 22 days, it was 9 mg/100 g in tuna after 20 days in ice (Perez-Villareal & Pozo 1990; Rehbein *et al.* 1994) TVB-N values of pangasius fillets were found to be within acceptable limits till 9 days storage in ice (Rao *et al.* 2013). TMA-N and TVB-N accumulated rapidly in sardines at ambient temperature compared to iced storage. Average shelf-life of 9.5 days was reported for sardines stored in ice (Ababouch *et al.* 1996). The TVB-N value of tilapia reached 38.75 mg/100 g after 15 days of ice storage (Adoga *et al.* 2010). There are considerable problems in maintaining quality during the period of post-catch handling and distribution of fish. The time lapse before icing and the exposure to ambient temperatures encourage proliferation of microorganisms, results in enhanced bacterial activity and induced enzymatic spoilage (Jeyasekaran *et al.* 2005). In a study, quality changes of Mediterranean hake were investigated during post catch handling and distribution and more rapid chemical deterioration was observed in seawater stored samples compared to ice storage (Venieri *et al.* 2013).

Scrombroid fish (tuna, albacore) contain characteristically high levels of free histidine in their muscle tissue, which can be converted to histamine by the action of bacterial decarboxylase. Bacteria known to be capable of decarboxylating histidine include *Vibrio, Proteus morganii* and *Klebsiella pneumoniae*, all having minimum growth temperatures between 8°C and 15°C. Since histamine is only produced at temperatures above

8°C, adequate temperature control is necessary to reduce or prevent its formation. Other biogenic amines formed as a result of bacterial activity on free amino acids in fish muscle are cadaverine and tyramine (Fraser & Sumar 1998). However, psychrotolerant histamine-producing bacteria can be found in chilled fish and fish products. Psychrotolerant bacteria *Morganella morganii* and *Photobacterium phosphoreum* were found in tuna stored at 2°C (Emborg *et al.* 2005). A significant increase in histamine, putrescine and cadaverine of barramundi slices (Bakar *et al.* 2010), sardine (Gökoğlu *et al.* 2004) were observed throughout the storage period at 4°C.

Since fish lipid is highly unsaturated, it is prone to oxidation (Church 1998). Both hydrolytic and oxidative rancidity should be considered during the storage of oily fish. During the advanced stages of lipid oxidation, the breakdown of hydroperoxides generates low molecular-weight carbonyl and alcohol compounds that could lead to the appearance of objectionable odours (Sikorski *et al.* 1990). Oxidation of the oil in oily fish gives rise to rancid odours and flavours; these can limit the storage life of such species more quickly than the protein changes that govern the extractable protein value.

An important stage in the oxidation is the reaction of oxygen with unsaturated fatty acid molecules to form hydroperoxides; the amount of these can be used as a measure of the extent of oxidation in the early stages, by using the peroxide test. An increase in the peroxide value (PV) is most useful as an index of the earlier stages of oxidation; as oxidation proceeds and peroxides are degraded, the PV can start to fall (Gökoğlu *et al.* 2012). Low initial PV in *Pangasius sutchi* reached 16.64 meq/kg after 20 days of ice storage (Hossain *et al.* 2005). The peroxides are presumed to be eventually further oxidized to aldehydes and ketones, which cause disagreeable rancid odours and taste. PVs of tilapia were under acceptable levels throughout storage in ice. Icing protected the fish against oxidation for 21 days (Adoga *et al.* 2010).

The thiobarbituric acid (TBA) test is used to detect secondary oxidation products, and an increase in TBA value is an indication of oxidative deterioration. An increase in TBA values of skipjack was observed during storage in ice (Mazorra-Manzano *et al.* 2000). Slurry ice slowed down the formation of thiobarbituric acid reactive substances in horse mackerel (Losada *et al.* 2005). TBA values of horse mackerel significantly increased up to 14 day of storage in ice (Aubourg 2001). Ice storage protected rohu against oxidation; low TBA values and limited oxidative rancidity were observed during the storage (Dhanapal *et al.* 2013).

The *para*-anisidine value (p-Av) is a guide for determining the decomposition products of hydroperoxides. The *p*-anisidine test provides useful information on non-volatile carbonyl compounds formed in oils during processing. It is often used to detect secondary oxidation products. Unsaturated aldehydes and high-molecular-weight decomposition products, such as triglyceride dimers and non-volatile parts of fatty acids that react with anisidine, are good indicators of lipid oxidation (Kolanowski *et al.* 2007; Laguerre *et al.* 2007). Good quality oil should have a p-Av of less than 2 (Subramanian *et al.* 2000).

Conjugated dienes are produced because of lipid oxidation and can only be formed from fatty acids with at least two double bonds. The non-conjugated double bonds are converted to conjugated double bonds, which absorb ultraviolet light strongly at 233 nm. UV absorption decreases with the degradation of CDs. The CDs provides a marker of the early stages of lipid peroxidation (Halliwell & Gutteridge 1985). Over 90% of hydroperoxides formed by lipoperoxidation have a conjugated dienic system resulting from stabilization of the radical state by double bond rearrangement (Gökoğlu *et al.* 2012). CD formation did not show any significant trend during the storage process for whole and filleted horse mackerel due to instability and capability interacting with other constituents (Aubourg 2001).

Free fatty acids (FFA) are formed through chemical- or enzyme-mediated hydrolysis of triacylglycerides and expressed as oleic acid per 100 g of oil. FFA determination is an important reaction that indicates the post-mortem changes occurring in fish lipids (Chaijan *et al.* 2006). The presence of FFA is due to hydrolysis of lipids and is undesirable since the fatty acids may be converted to odorous volatiles such as carbonyls, alkenes and alcohols (Karungi *et al.* 2004). FFA content provided an increasing trend in salmon muscle during chilled storage (Aubourg *et al.* 2005). In whole horse mackerel samples, lipid hydrolysis increased gradually during chilled storage; a significant increase was observed at day 6 of ice storage compared to the control sample, and the FFA content at the end of the experiment (day 19) reached 5.3% (Aubourg 2001).

5.3 Microbiological changes

Bacterial growth is the main cause of fish spoilage; therefore, it is logical to use bacteria numbers as an index of quality. For high quality fresh fish, the number of bacteria present on the surface varies from 3 to 4 $\log_{10}$ colony-forming units (CFU)/g. On gills, counts are normally 1 or 2 orders higher, and intestinal counts can reach 9 $\log_{10}$ CFU/g (Sikorski *et al.* 1990). The spoilage of fish when stored in ice is mainly a result of psychrotrophic bacteria which easily grow and multiply at low temperatures (Shewan 1977; Lima dos Santos *et al.* 1981; Surendran & Gopakumar 1981). The storage life of iced fish is usually affected by initial microbial load of the fish and storage temperature (Church 1998). Advantages of the ice preservation are maximum possibility of preserving the natural nutritional and functional properties of the fish (Dhanapal *et al.* 2013). *Psychrotrophic* bacteria belonging to *Pseudomonas* spp. and *Shewanella putrefaciens* dominate the spoilage flora of iced stored fish. Differences exist in the spoilage profile depending on the dominating bacterial species. *Shewanella* spoilage is characterized by TMA and sulphides (H_2S), whereas *Pseudomonas* spoilage is characterized by absence of these compounds and occurrence of sweet, rotten sulphydryl odours. As this is not typical of temperate, marine fish species which have been widely studied, this may explain the hypothesis that bacteria are not involved in the spoilage process of tropical fish (Huss 1995).

The counts of psychrophiles, sulphur-producing bacteria, *Pseudomonas* and *Aeromonas*s pp. increased during ice storage of rohu (Dhanapal *et al.* 2013).

5.4 Enzymatic changes

Numbers of proteolytic enzymes are found in muscle and viscera of the fish after catch. These enzymes contribute to post mortem degradation in fish muscle and fish products during storage and processing peptides and free amino acids can be produced as a result of autolysis of fish muscle proteins, which lead towards the spoilage of fish meat as an outcome of microbial growth and production of biogenic amines Belly bursting is caused by leakage of proteolytic enzymes from pyloric caeca and intestine to the ventral muscle (Ghaly *et al.* 2010).

The initial quality loss is mainly explained by autolytic changes, such as degradation of nucleotides (ATP-related compounds) by autolytic enzymes. The loss of the intermediate nucleotide, inosine monophosphate (IMP), is responsible for the loss of fresh fish flavour (Huss 1995). These autolytic changes make catabolites available for bacterial growth. Post-mortem degradation of ATP in fish muscle occurs due to endogenous enzymatic activity. This degradation goes through the intermediate products adenosine diphosphate (ADP), adenosine monophosphate (AMP), IMP, inosine (INO), and Hx (Perez-Villareal & Pozo 1990). The adenosine nucleotides and their catabolites have been used as indicators of freshness in several species; however, differences in the nucleotide-degradation patterns from species to species have been reported (Ohashi *et al.* 1991; Murata and Sakaguchi 1986). ATP and ADP were not detected in flounder during storage in ice. A decrease of 55% in IMP concentration was observed during the first 2 days of storage. Hypoxanthine increased up to the 7th day (Massa *et al.* 2005). The K value (percentage ratio of INO and Hx to the sum of ATP and all products of ATP degradation) reached 60% at 25 days of iced storage in rohu fish (*Labeo rohita*) (Mohan *et al.* 2006)

Post-mortem tenderization is one of the most important quality attributes of fish muscle. Loss of freshness is due to a complex combination of biochemical, chemical and physical processes, and is followed by muscle spoilage due to microbiological contamination. However, the first changes occurring in post-mortem fish muscle are due to endogenous enzymes promoting proteolysis of muscle proteins and connective tissue as well as fat hydrolysis. Indeed, the muscle is not significantly contaminated by bacteria at this stage. Tenderization is enzymatic in nature; physicochemical conditions (pH, osmotic pressure) may modulate the proteolytic action of endogenous enzymes (Delbarre-Ladrat *et al.* 2006).

Spoilage microflora produces enzymes that cause proteolysis, deamination, and decarboxylation, resulting in accumulation of unpleasant metabolites and loss of taste substances (Huss 1988; Sikorski *et al.* 1990).

Autolytic changes in chilled or frozen or frozen fish are summarized in Table 5.1.

Table 5.1 Summary of autolytic changes in chilled or frozen fish*

Enzyme (s)	Substrate	Changes encountered	Prevention
Glycolytic enzymes	Glycogen	Production of lactic acid, pH of tissue drops, loss of water-holding capacity in muscle high temperature rigor may result in gaping	Fish should be allowed to pass through rigor at temperatures as close to 0°C as practically possible pre-rigor stress must be avoided
Autolytic enzymes involved in nucleotide breakdown	ATP ADP AMP IMP	Loss of fresh fish flavour, gradual production of bitterness with Hx* (later stages)	Same as above – rough handling or crushing accelerates breakdown
Cathepsins	Proteins, peptides	Softening of tissue making processing difficult or impossible	Avoid rough handling during storage and discharge
Chymotrypsin, trypsin, carboxypeptidases	Proteins, peptides	Autolysis of visceral cavity in pelagic (belly-bursting)	Problem increased with freezing/thawing or long-term chill storage
Calpain	Myofibrillar proteins	Softening molt-induced softening, in crustaceans	Removal of calcium thus preventing Activation
Collagenases	Connective tissue	Gaping of fillets softening of muscle tissue	Connective tissue degradation related to time and temperature of chilled storage
TMAO demethylase	TMAO	Formaldehyde-induced toughening of frozen gadoid fish	Store fish at temperatures less than or equal to 30°C physical abuse and freezing/thawing accelerate formaldehyde-induced toughening

Food and Agriculture Organization of the United Nations 2005. Post-harvest changes in fish. <http://www.fao.org/fishery/topic/12320/en> Reproduced with permission.

5.5 Physical changes

Texture is an extremely important property of fish muscle, whether raw or cooked. Fish muscle may become tough as a result of frozen storage or soft and mushy as a result of autolytic degradation. It has been confirmed that storage temperature during handling and operating process generally has a distinctive effect on fish texture measurement. The most common texture defects are muscle softening and gaping, caused by pre- and post-mortem treatment. Existing problems are mostly associated with the changes of chemical compositions and the degradation of muscle proteins. There are other numerous interacting factors along with physical factors such as species, age, size, feeding ingredients, sample heterogeneity and gaping; chemical factors such as water and fat contents and distributions, and collagen contents; and diverse treatments such as storage time and temperature, freezing, chilling high pressure processing, salting and smoking. Among texture attributes firmness, also termed as hardness – an essential evaluating parameter of fish freshness – is closely associated with the human visible acceptability of fish products (Cheng *et al.* 2014). Stiffness of rohu fish (*Labeo rohita*) decreased during the storage in the ice-box covered with ice (Jain *et al.* 2007).

Changes in surface coloration of fish and shellfish, as well as the altered colour of the flesh, result mainly from enzymatic and non-enzymatic oxidation. The yellow, orange and red colour, or colourlessness, of fish and shellfish is caused by oxidation of carotenoids present in large amounts in the skin, shells or exoskeletons. The dark-brown to black pigmentation of fish skin, induced by melanins, fades away, the skin becoming lustreless and losses its iridescent appearance. Due to chemical enzymatic oxidation of heme pigments, dark red muscles become brown. The flesh of fresh fish is translucent, but, in stale fish it tends to be opaque. Slime on the skin, initially watery and clear, becomes cloudy, clotted, and discoloured as a result of increased bacterial activity (Sikorski *et al.* 1990).

Chemical groups affecting fish flesh colour are hemes, carotenoids and melanins. Colour of fish is affected by different constituents in whole fish; greyness arises from melanins and red or red/brown from blood and dark muscles (Hutching 1994). Pigments in darker meat are especially vulnerable to oxidation, which causes deep yellow or brown discoloration during handling, chilling, and frozen storage (Suvanich *et al.* 2000).

5.6 Sensory changes

Among sensory changes which include the smell, appearance, texture and taste of fish meat, the most obvious is the appearance of rigor mortis, and it depends on many factors, such as the temperature of water and storage, handling with the fish, size and physical condition of the fish, and the method of paralysing or catch (Bojanic *et al.* 2009). Sensory characteristics of catfish were excellent at the first 3 days of iced storage; however, there was a noticeable decrease in the attributes from day 10. The eyes were sunken, the gills were red brown with mucus, the skin lost some brightness and texture According to results, fresh catfish samples kept very well in ice and in good condition up to day 17 (Adeosun *et al.* 2014). Sensory scores of iced rohu decreased gradually with increasing storage time (Dhanapal *et al.* 2013). Tilapia samples were found in acceptable conditions after 15 days in ice storage. Rejection of raw fish by the taste panellists was mainly characterized by strong fishy to sour odours and soft texture (Adoga *et al.* 2010). Similar shelf-life for tilapia stored in ice was reported in the other research (Kapute *et al.* 2012). A higher sensory shelf-life (15 days) was observed for horse mackerel treated under slurry icing, when compared with flake iced fish (5 days) (Losada *et al.* 2005).

References

Ababouch, H., Souibri, L., Rhaliby, K., Ouahdi, O., Battal, M. & Busta, F.F. (1996). Quality changes in sardines (*Sardina pilchardus*) stored in ice and at ambient temperature. *Food Microbiology*, **13**:123–132.

Adeosun, O., Olukunle, O. & Akande, G.R. (2014).Proximate composition and quality aspects of iced wild and pond-raised African catfish (*Clarias gariepinus*). *International Journal of Fisheries and Aquaculture*, **6**(3): 32–38.

Adoga, I.J., Joseph, E. & Samuel, O.F. (2010). Storage life of tilapia (*Oreochromis niloticus*) in ice and ambient temperature. *Researcher*, **2**(5): 39–44.

Ashie, I.N.A., Smith, J.P. & Simpson, B.K. (1996). Spoilage and shelf-life extension of fresh fish and shellfish. *Critical Reviews in Food Science and Technology*, **36**: 87–121.

Aubourg, S.P. (2001). Damage detection in horse mackerel (*Trachurus trachurus*) during chilled storage. *Journal of American Oil Chemists' Society*, **78**: 857–862.

Aubourg, S.A., Vinagre, J., Rodríguez, A., *et al.* (2005). Rancidity development during the chilled storage of farmed Coho salmon (*Oncorhynchus kisutch*). *European Lipid Science and Technology*, **107**(6): 411–417.

Bakar, J., Yassoralipour, A., Bakar, F.A. & Rahman, R.A. (2010). Biogenic amine changes in barramundi (*Lates calcarifer*) slices stored at 0°C and 4°C. *Food Chemistry*, **119**: 467–470.

Bojanic, K., Kozacinski, L., Filipovic, I., Cvrtila, Z., Zdolec, N. & Njari B. (2009). Quality of seabass meat during storage on ice. *MESO*, **11**(1): 74–80.

Chaijan, M., Benjakul, S., Visessanguan, W. & Faustman, C. (2006). Changes of lipids in sardine (*Sardinella gibbosa*) muscle during iced storage. *Food Chemistry*, **99**: 83–91.

Cheng, J.H., Sun, D.W., Hang, Z. & Zeng, X.A. (2014). Texture and structure measurements and analyses for evaluation of fish and fillet freshness quality: A review. *Comprehensive Reviews in Food Science and Food Safety*, **13**: 52–61.

Church, N. (1998). MAP fish and crustacean sensory enhancement. *Food Science and Technology Today*, **12**(2): 73–83.

Connell, J.J. (1995). Quality deterioration and extrinsic quality defects in raw material. In: *Control of Fish Quality* (Ed. Connell, J.J.), 4th edn, pp. 31–35, Fishing News Books Ltd., Oxford.

Dagbjartsson, B. (1975). Utilization of blue whiting for human consumption. *Journal of the Fisheries Research Board of Canada*, **32**: 747–751.

Delbarre-Ladrat, C., Chéret, R., Taylor, R., & Verrez-Bagnis, V. (2006).Trends in postmortem aging in fish: understanding of proteolysis and disorganization of the myofibrillar structure. *Critical Reviews in Food Science and Nutrition*, **46** (5): 409–421.

Dhanapal, K., Sravani, K., Balasubramanian, A. & Reddy, G.V.S. (2013). Quality determination of rohu (*Labeo rohita*) during ice storage. *Tamilnadu Journal of Veterinary & Animal Sciences*, **9** (2): 146–152.

Emborg, J., Laursen, B.G. & Dalgaard, P. (2005).Significant histamine formation in tuna (*Thunnus albacares*) at 2–8C – effect of vacuum- and modified atmosphere-packaging on psychrotolerant bacteria. *International Journal of Food Microbiology*, **101**: 263–279.

FAO. (2005). Post-harvest changes in fish. In: *FAO Fisheries and Aquaculture Department*. Food and Agriculture Organization, Rome, Italy.

Fraser, O.P. & Sumar, S. (1998). Compositional changes and spoilage in fish (part II) -Microbiological induced deterioration. *Nutrition & Food Science*, **6**: 325–329.

Ghaly, A.E., Dave, D., Budge, S. & Brooks, M.S. (2010). Fish spoilage mechanisms and preservation techniques: Review. *American Journal of Applied Sciences*, **7** (7): 859–877.

Gökoğlu, N., Yerlikaya, P., & Cengiz, E. (2004). Changes in biogenic amine contents and sensory quality of sardine (*Sardina pilchardus*) stored at 4°C and 20°C. *Journal of Food Quality*, **27** (3): 221–231.

Gökoğlu, N., Yerlikaya, P., & Topuz, O.K. (2012). Effects of tomato and garlic extracts on oxidative stability in marinated anchovy. *Journal of Food processing and Preservation*, **36**(3): 191–197.

Hardy, R. (1980). Fish lipids. Part 2. In: *Advances in Fish Science and Technology*, pp. 103–110. Fishing News Books, Oxford.

Halliwell, B. & Gutteridge, J.M.C. (1985). The chemistry of oxygen radicals and other oxygen derived species. In: *Free Radicals in Biology and Medicine* (Eds Halliwell, B. & Gutteridge, J.M.C.), 4th edn, pp. 20–64, Oxford University Press, New York.

Hossain, M.I., Islam, M.S., Shikha, F.K., Kamal, M. & Islam, M.N. (2005). Physicochemical changes in Thai pangas (*Pangasius sutchi*) muscle during ice-storage in an insulated box. *Pakistan Journal of Biological Sciences*, **8** (6): 798–804.

Huss, H.H. (1988). *Fresh fish quality and quality changes.* FAO/DANIDA.A Training Manual, FAO Fisheries Series No. 29, FAO, Rome, Italy.

Hutching J.B. (1994). *Food Color and Appearance*. Blackie Academic & Professional, New York.

Jain, D., Pathare, P.B. & Manikantan, M.R. (2007). Evaluation of texture parameters of rohu fish (*Labeo rohita*) during iced storage. *Journal of Food Engineering*, **81**: 336–340.

Jeyasekaran, G.,Maheswari, K., Ganesan, P., Shakila, R.J. & Sukumar, D. (2005). Quality changes in ice-stored tropical wire-netting reef cod (*Epinephelus merra*). *Journal of Food Processing and Preservation*, **29**: 165–182.

Kapute, F., Likongwe, J.S., Kang'ombe, J., Mfitilodze, B. & Kiiyukia, C. (2012). Shelf-life of whole Lake Malawi Tilapia (Chambo) stored in ice. Third RUFORUM Biennial Meeting 24 - 28 September 2012, Entebbe, Uganda, pp. 461–465.

Karungi, C., Byaruhanga, Y. B. & Muyonga, J. H. (2004). Effect of pre-icing duration on quality deterioration of iced Nile perch (*Lates niloticus*). *Food Chemistry*, **85**(1): 13–17.

Kolanowski, W., Jaworska, D. & Weissbrodt, J. (2007). Importance of instrumental and sensory analysis in the assessment of oxidative deterioration of omega-3 long-chain polyunsaturated fatty acid-rich foods. *Journal of the Science of Food and Agriculture*, **87**: 181–191.

Laguerre, M., Lacomte, J. & Villeneuve, P. (2007). Evaluation of the ability of antioxidants to counteract lipid oxidation: Existing methods, new trend and challenges. *Progress in Lipid Research*, **46**: 244–282.

Lima Dos Santos, C.A.M., James, D. & Teutscher, F. (1981). *Guidelines for chilled fish storage experiments.* FAO Fisheries Technical Paper **210**: 17–21.

Losada, V. Piñeiro, C., Barros-Velázquez, J. & Aubourg, S. (2005). Inhibition of chemical changes related to freshness loss during storage of horse mackerel (*Trachurus trachurus*) in slurry ice. *Food Chemistry*, **93**: 619–625.

Massa A.E., Palacios, D.L., Paredi, M.E. & Crupkin, M. (2005). Postmortem changes in quality indices ofice-stored flounder (*Paralichthys patagonicus*). *Journal of Food Biochemistry*, **29**: 570–590.

Mazorra-Manzano, M.A., Pacheco-Aguilar, V.R., Díaz-Rojas, E.I. & Lugo-Sánchez, M.E. (2000). Postmortem changes in black skipjack muscle during storage in ice. *Journal of Food Science*, **65**(5): 774–779.

Mohan, M., Ramachandran, D. & Sankar, T.V. (2006). Functional properties of rohu (*Labeorohita*) proteins during iced storage. *Food Research International*, **39** (8): 847–854.

Murata, M. & Sakaguchi, M. (1986). Storage of yellowtail (*Seriola quinqueradiata*) white and dark muscles in ice: changes in content of adenine nucleotides and related compounds. *Journal of Food Science*, **51**: 321–326.

Ohashi, E., Okamoto, M., Ozawa, A. & Fujita, T. (1991). Characterization of common squid using several freshness indicators. *Journal of Food Science*, **56**:161–163, 174.

Odoli, C.O. (2009). Optimal storage conditions for fresh farmed tilapia (Oreochromis niloticus) fillets. Thesis submitted in partial fulfilment of the requirements for the Degree of masters in science. Department of Food Science and Nutrition, University of Iceland.

Perez-Villarreal, B. & Pozo, R. (1990).Chemical composition and ice spoilage of albacore. *Journal of Food Science*, **55**: 678–682.

Rao, B.M., Murthy, L.N. & Prasad, M.M. (2013). Shelf-life of chill stored pangasius (*Pangasianodon hypophthalmus*) fish fillets: effect of vacuum and polyphosphate. *Indian Journal of Fisheries*, **60**(4): 93–98.

Rehbein, H., Martinsdottir, E., Blomsterberg, F., Valdimarsson, G. & Oehlenschlaeger, J. (1994). Shelf-life of ice-stored redfish, *Sebastes marinus and S. mentella*. *International Journal of Food Science and Technology*, **29**: 303–313.

Shawyer, M. & Pizalli, A.F.M. (2003). The use of ice on small fishing vessels. FAO Fisheries Technical Paper No 436. FAO, Rome, Italy.

Shewan, J.M. (1977). The bacteriology of fresh and spoiling fish and the biochemical changes induced by bacterial action. In: *Handling, Processing and Marketing of Tropical Fish* (Eds P. Sutchliffe & J. Disney), pp. 51–66, Tropical Products Institute, London.

Sikorski Z.E. & Kolakowski E. (2000). Endogenous enzyme activity and seafood quality: influence of chilling, freezing, and other environmental factors. In: *Seafood Enzymes: Utilization and Influence on Postharvest Seafood Quality* (Eds N.F. Haard & B.K. Simpson), pp. 451–487, Marcel Dekker, New York.

Sikorski, E.Z., Kolakowska, A. & Burt, J.R. (1990). Postharvest biochemical and microbial changes. *Seafood: Resources, Nutritional Composition, and Preservation* (Ed. Sikorski, Z.E.), pp. 56–72, CRC Press Inc., Boca Raton, FL.

Sivertsvik, M., Jeksrud, W.K. & Rosnes, J.T. (2002). A review of modified atmosphere packaging of fish and fishery products-significance of microbial growth, activities and safety. *International Journal of Food Science and Technology*, **37**: 107–127.

Subramanian, R., Nandini, K.E., Sheila, P. M., *et al.* (2000). Membrane processing of used frying oils. *Journal of the American Oil Chemists Society*, **77**: 323–328.

Surendran, P.K. & Gopakumar, K. (1981). Selection of bacterial flora in the chlortetracycline treated oil sardine (*Sardinella longiceps*), Indian mackerel (*Rastrelliger kanagurta*) and prawn (*Metapenaeus dobsoni*) during ice storage. *Fishery Technology*, **18**: 133–141.

Suvanich, V., Marshall, D.I. & Jahncke, M.I. (2000). Microbiological and colour quality changes of channel catfish frame mince during chilled and frozen storage. *Journal of Food Science*, **65**(1): 151–154.

Venieri, D., Theodoropoulos, C., Lagkadinou, M. & Iliopoulou-Georgudaki, J. (2013). Effects of ice and seawater storing conditions on the sensory, chemical and microbiological quality of the Mediterranean hake (*Merluccius merluccius*) during post-catch handling and distribution. *International Journal of Agricultural, Biosystems Science and Engineering*, **7** (2): 46–51.

Whittle, K.J., Hardy, R. & Hobbs, G. (1990). Chilled fish and fishery products. In: *Chilled Foods* (Ed. Gormley, T.R.), pp. 87–116, Elsevier Science Publishers, London.

CHAPTER 6

Refrigeration

6.1 Introduction

An alternative cooling agent to ice is refrigerated air. Refrigeration is a process of removing heat from a confined area or from a product for the purpose of lowering the temperature of the area or product. There are two aspects of refrigeration, which may be defined as cooled (or chilled and frozen). Chilled storage implies temperature reduction to some point above the freezing temperature of the product. Chilled conditions for aquatic food products will include temperatures down to about −1°C, the initial freezing temperature of many aquatic products. The freezing process involves some change of state, usually of water, in temperature reduction process. Thus, freezing process require lowering temperatures to below the freezing point (Wheaton & Lawson 1985).

The first mechanical refrigerators for the production of ice appeared around the year 1860. In 1880 the first ammonia compressors and insulated cold stores were put into use in the USA. Electricity began to play a part at the beginning of the twentieth century and mechanical refrigeration plants became common in some fields: breweries, slaughter-houses, fishery and ice production, for example. After the

Seafood Chilling, Refrigeration and Freezing: Science and Technology, First Edition.
Nalan Gökoğlu and Pınar Yerlikaya.
© 2015 John Wiley & Sons, Ltd. Published 2015 by John Wiley & Sons, Ltd.

Second World War the development of small hermetic refrigeration compressors evolved, and refrigerators and freezers began to take their place in the home. Today, these appliances are regarded as normal household necessities.

Cold air passed over the surface of a fish will rapidly cool it. In a chill room, heat from the fish will warm the air around it. The warm air rises and is cooled by the refrigeration system. This cold air then falls or is blown by fans, back to the fish surface. Good circulation of air is necessary to maintain uniform temperatures in the fish in store. Unfortunately 10 000 times less heat is required to warm a given volume of air from 0 to 0.5°C, than to warm the same volume of crushed ice. Thus, compared with ice, large refrigeration systems and larger volumes of air are needed to cool a given weight of fish. Cooling with cold air is therefore generally more inefficient and more expensive than cooling with ice (Warren 1986).

Fish that are cooled in cold air soon become dry. This is because the air removes moisture from the fish surface, causing loss of weight and loss of eating quality. The water travels with the air to the cooling coils where it is deposited as ice (frost). This can interfere with the cooling of the air, as it acts as an insulating layer. The evaporator must be regularly defrosted to prevent this. Fish stacked high up in the chill room, close to the evaporator, will receive the coldest air. If the average store temperature is set at 1°C, the air temperature at the evaporator will be less than 0°C and the fish here may freeze. Often chill rooms are used together with ice, which helps to slow down the speed at which the ice melts. It is importance remember that, for tee to cool effectively. It must be allowed to melt and the chill room temperature should not fall below 2°C to 3°C (Warren 1986).

6.2 Fundamentals of refrigeration

Refrigeration is the action of removing heat from an enclosed space or material for the purpose of lowering its temperature. A refrigeration system must provide a means by which heat can move away. Refrigeration systems do this by providing a cold surface near the material to be cooled. This surface, colder than the material, causes heat to transfer from that material through the cold surface. Because heat only flows from a warmer body to a colder body, the temperature of the cold surface must be less than that of the refrigerated material. A refrigeration system is the collection of equipment that generates cold and hot surface stopper form refrigeration. Once heat or thermal energy passes through the cold surface, the objective of the refrigeration system is to transport the energy to another location (Fenton 2010).

There are several fundamental terms. Before an introduction to refrigeration systems these terms should be considered.

- *Temperature*: Temperature is a very central property in refrigeration. Almost all refrigeration systems have the purpose of reducing the temperature of a substance like the air in a room or the objects stored in that room. The SI-unit for temperature *Kelvin* [K] is an absolute temperature because its reference point (0 K) is the lowest temperature that it in theory would be able to obtain. When working with refrigeration systems the temperature unit *degree Celsius* (°C) is a more practical unit to use. Celsius is not an absolute temperature scale because its reference point (0°C) is defined by the freezing point of water. The only difference between Kelvin and Celsius is the difference in reference point (Danfoss 2007).
- *Force and pressure*: The SI-unit for force is *Newton* (N; kg m/s^2). A man wearing skis can stand in deep snow without sinking very deep but if he steps out of his skis his feet will probably sink very deep into the snow. In the first case the weight of the man is distributed over a large surface (the skis). In the second case the same weight is distributed on the area of his shoe soles, which is a much smaller area than the area of the skis. The difference between these two cases is the pressure that the man exerts on the snow surface. Pressure is defined as the force exerted on an area divided by the size of the area. In the example with the skier, the force (gravity) is the same in both cases but the areas are different. In the first case the area is large and so the pressure becomes low. In the second case the area is small and so the pressure becomes high. In refrigeration pressure is mostly associated with the fluids used as refrigerants. When a substance in liquid or vapour form is kept within a closed container the vapour will exert a force on the inside of the container walls. The force of the vapour on the inner surface divided by its area is called the *absolute* pressure. The pressure above atmospheric pressure is also often referred to as *gauge pressure* (Danfoss 2007).

- *Heat, work, energy and power*: Heat and work are both forms of energy that can be transferred between objects or systems. The transfer of heat is closely connected to the temperature (or temperature difference) that exists between two or more objects. By itself, heat is always transferred from an object with high temperature to objects with lower temperatures. Heating water in a pot on a stove is a good everyday example of the transfer of heat. The stove plate becomes hot and heat is transferred from the plate through the bottom of the pot and to the water. The transfer of heat to the water causes the temperature of the water to rise. The methods of transfer between objects are different but for a process with both heat and work transfer it is the sum of the heat and work transfer that determines the outcome of the process. The SI-unit *Joule* (J) is used to quantify energy, heat and work. The amount of energy needed to increase the temperature of 1 kg of water from 15 to 16°C is 4.187 kJ. The 4.178 kJ can be transferred as heat or as work – but heat would be the most used practical solution in this case. There are differences in how much energy is required to increase the temperature of various substances by 1 K. For 1 kg of pure iron approximately 0.447 kJ is needed, whereas for 1 kg of atmospheric air only app. 1.0 kJ is needed. The property that makes the iron and air different with respect to the energy needed for causing a temperature increase is called the 'specific heat capacity'. It is defined as the energy required causing a temperature increase of 1 K for 1 kg of the substance. The unit for specific heat capacity is J/kg K. The rate at which energy is transferred is called power (Danfoss 2007).
- *Substances and phase changes*: All substances can exist in three different phases: solid, liquid and vapour. Water is the most natural example of a substance that we use almost every day in all three phases. The solid form we call ice, the liquid form we just call water, and the vapour form we call steam. What is common to these three phases is that the water molecules remain unchanged, meaning that ice, water and steam all have the same chemical formula: H_2O. When taking a substance in the solid to the liquid phase the transition process is called melting and when taking it further to the vapour phase the transition process is called boiling (evaporation). When going in the opposite direction taking a substance from the vapour to the liquid phase the transition process is called condensing and when taking it

further to the solid phase the transition process is called freezing (solidification). At constant pressure the transition processes display a very significant characteristic. When ice is heated at 1 bar its temperature starts rising until it reach 0°C then the ice starts melting. During the melting process the temperature does not change all the energy transferred to the mixture of ice and water goes into melting the ice and not into heating the water. Only when the ice has been melted completely will the further transfer of energy cause its temperature to rise. The same type of behaviour can be observed when water is heated in an open pot. The water temperature increases until it reaches 100°C then evaporation starts. During the evaporation process the temperature remains at 100°C. When all the liquid water has evaporated the temperature of the steam left in the pot will start rising. The temperature and pressure a substance is exposed to determine whether it exists in solid, liquid, or vapour form – or in two or all three forms at the same time. In our local environment iron appears in its solid form, water in its liquid and gas forms, and air in its vapour form (Danfoss 2007).

- *Latent heat*: Latent, or hidden, heat is the heat energy absorbed or released when a substance changes state from solid to liquid (or vice versa) or from liquid to gas. Latent heat transfer is not accompanied by a temperature change but is far more important in refrigeration than the transfer of sensible heat because the quantities of energy involved are greater. When water at 0°C changes state to become ice at 0°C, the latent heat of fusion, or solidification, which for water is 335 kJ/kg is released to the surrounding environment. Conversely, the same amount of energy is absorbed when ice at 0°C melts to become water at 0°C. As noted above, the amount of heat then required to rise the temperature of this water is only 4.187 kJ/kg/°C, hence as much heat is required to change 1 kg of ice into 1 kg of water as is required to rise the temperature of that same 1 kg of water from 0°C to 85°C. These principles are the basis of conventional refrigeration cycles. Normally, the refrigerant is brought into proximity with the product to be cooled, and there encouraged to undergo a change of state from liquid to gas, thus absorbing its latent heat of evaporation from the product. It is then transported away from the product and forced to condense to a liquid state, during which process it releases the heat it has absorbed to the environment. The cooling properties of a substance which changes state are far greater than those of one which does not (Preston & Vincent 1985).

When looking at the boiling process of water the energy needed for evaporating 1 kg of water is 2501 kJ. The refrigeration effect in refrigeration systems is based on the use and control of the phase transition processes of evaporation. As the refrigerant evaporates it absorbs energy (heat) from its surroundings and by placing an object in thermal contact with the evaporating refrigerant it can be cooled to low temperature (Danfoss 2007).

- *Superheat*: Superheat is a very important term in the terminology of refrigeration but it is unfortunately used in different ways. It can be used to describe a process where refrigerant vapour is heated from its saturated condition to a condition at higher temperature. The term superheat can also be used to describe – or quantify – the end condition of the before-mentioned process. Superheat can be quantified as a temperature difference between the temperature measured with a thermometer and the saturation temperature of the refrigerant measured with a pressure gauge. Therefore, superheat cannot be determined from a single measurement of temperature alone – a measurement of pressure or saturation temperature is also needed. When superheat is quantified it should be quantified as a temperature difference and, consequently, be associated with the unit K. If quantified in °C it can be the cause of mistakes where the measured temperature is taken for the superheat, or vice versa. The evaporation process in a refrigeration system is one of the processes where the term superheat is used (Danfoss 2007).

6.3 Refrigeration systems

Thermodynamically, refrigeration may be produced by a number of different processes. Three different thermodynamic effects are used commercially to produce refrigeration, namely: the *vapour compression cycle*, the *absorption cycle* and the *Peltier effect*. The basic principle of the most common type of mechanical refrigeration is a cyclic thermodynamic process known as the *Rankine cycle* or a *vapour compression cycle*. The cycle consists of four sections (Figure 6.1):

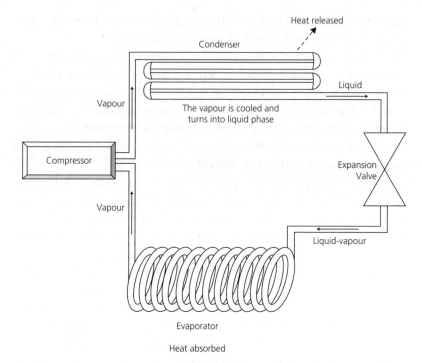

Figure 6.1 Schematic diagram of refrigeration system (adapted from Rao 2010).

1 *Compression*: saturated vapour at pressure P1 is compressed to pressure P2. Ideally, isoenthropic (adiabatic and reversible) compression is assumed. Mechanical work is supplied to the *compressor*.
2 *Condensation*: the compressed vapour is cooled until completely condensed as a saturated liquid. Ideally, cooling is assumed to take place at constant pressure. The heat removed from the condensed vapour is transferred to a cooling medium such as air or water. Physically, this step takes place in a heat exchanger serving as a *condenser*.
3 *Expansion*: the pressure of the liquid is released through a throttling element. The throttling process is supposed to be isenthalpic, not involving any exchange of energy.
4 *Evaporation*: heat is transferred to the liquid–vapour mixture until all the liquid is evaporated. This is the step of the cycle where useful refrigeration is generated.
  Physically, this last step occurs in a heat exchanger known as the *evaporator* or *diffuser*.

The fluid undergoing the cycle is called a *refrigerant*. Although, theoretically, the reverse Rankine cycle could be run with any fluid, only certain compounds and mixtures are suitable for use in practical refrigeration (Berk 2009). Several types of refrigeration systems are available, and each accomplishes the same task of removing heat from a space or material. The material may be a gas, liquid or solid. For example, the refrigeration system in a domestic refrigerator cools the inside air to a temperature less than the surroundings, thus maintaining an appropriate temperature for the storage of food. The means by which the cold surface is generated in the refrigerated space distinguishes one refrigeration system from another (Fenton 2010).

6.3.1 Vapour-compression system
The vapour-compression system is the most common refrigeration system currently in use. A vapour-compression system passes a fluid called a *refrigerant* through four components in sequence: evaporator, compressor, condenser, and expansion device (Fenton 2010). A refrigerant fluid in vapour state is compressed to a higher pressure and consequently a higher temperature. The higher-temperature gas is cooled and liquefied in a condenser. The cool liquid then passes through a restrictor to a lower-pressure area, cooling further in the process. The cold liquid can then be used to extract heat from a storage space or cooling area, this heat vaporising the cold, low pressure liquid in an evaporator. The cold vapour is then fed back to compressor to complete the cycle (Stringer and Dennis 2000). Since the fluid circulates through the system and returns to its original position, the system is also referred to as a *cycle*. The evaporator is the component where the actual cooling occurs. A mixture of liquid and vapour refrigerant at low pressure and low temperature enters the evaporator. The temperature of the refrigerant mixture in the evaporator is somewhat less than the temperature of the refrigerated space or material. Therefore, the evaporator absorbs heat from the space or material, causing the liquid refrigerant to boil or vaporize. The heat transported from the refrigerated space or material causes the space to maintain the needed temperature. The low-pressure, low-temperature refrigerant vapour leaves the evaporator and enters the compressor; upon exiting the compressor, the pressure and temperature are increased to that within the condenser. The condenser transfers heat from the high-pressure,

high-temperature refrigerant vapour to the surroundings, causing condensation of the vapour into a liquid. The liquid refrigerant flows to the expansion device, which decreases the pressure and causes the formation of a low-temperature mixture of vapour and liquid that enters the evaporator. In the vapour-compression system, the refrigerant circulates in sequence among the four components, which constitutes the vapour-compression cycle. An important feature of the cycle is the conversion of the refrigerant from liquid to vapour and back again as it moves through the system. In this way, the refrigerant is not consumed during operation of the system. All components of the system operate continuously, because the refrigerant is flowing steadily through the system (Fenton 2010).

The principal physical components of the vapour compression refrigeration machine are as follows.

The *compressor*: A refrigeration compressor is a pump that adds energy to a refrigeration system. In a refrigeration cycle, the compressor has two main functions within the refrigeration cycle. One function is to pump the refrigerant vapour from the evaporator so that the desired temperature and pressure can be maintained in the evaporator. The second function is to increase the pressure of the refrigerant vapour through the process of compression, and simultaneously increase the temperature of the refrigerant vapour (Dincer & Kanoglu 2010). Energy must be added in order to reject heat to the condenser cooling fluid. Compressors may be classified in several ways. One method, which relies on sealing characteristics of the compressor, is semihermetic and hermetic classifications.

Hermetic compressors are sealed units containing all compressor parts and the drive motor. These units cannot be dissembled for repair. Thus, failure of any compressor or motor part requires replacement of the entire unit, a process requiring the refrigeration lines to be opened. These units are usually found on household refrigerators, air conditioners and similar appliances.

Semihermetic compressors (single- or double-acting) were developed to avoid the disadvantages of the hermetic compressors. Semihermetic compressors are identical to the hermetic types, but the motor and compressor are constructed in a fabricated enclosure with bolted sections or access panels to facilitate servicing. These compressors are manufactured in small and medium capacities and their motor capacities can reach 300 kW. For this reason they are cheap and another advantage is that they are compact. Also, they do not have a leakage problem (Dincer & Kanoglu 2010).

Compressors may be also classified as reciprocating, rotary, centrifugal or jet types. Centrifugal compressors are typically used in large installations requiring 50 to 100 tonnes or more of refrigeration capacity. Jet compressors typically are used for air conditioning and similar uses where temperatures below 0°C are not normally necessary. Thus, they are rarely used for frozen storage of aquatic products. Rotary and reciprocating compressors are both used for refrigeration of aquatic products. Reciprocating piston compressors operate by low-pressure refrigerant in on the down stroke of the piston. As the piston moves upward the refrigerant is trapped in the space above the piston and compressed sufficiently to force it out through the exhaust valve into the high pressure side of the system. Since the compression is rapid the heat generated during compression does not have time to transfer out through the cylinder walls. Thus, both pressure and temperature of exhausted refrigerant are above that of the intake side of the refrigerant.

Reciprocating compressors are manufactured with very small clearances between the piston at the top of its intake and top of the cylinder chamber. Since liquid refrigerant is essentially incompressible, any liquid entering the compressor may seriously damage it. Refrigerant enters many compressors through the crankcase. Lubricating oil in the crankcase is picked up by the refrigerant and carried into the piston and out into the refrigeration system. The oil in the refrigerant lubricates the piston, a necessary function to protect the compressor. However, use the wrong oil or failure to provide proper oil tapping can deplete the compressor oil supply and lead to severe compressor damage (Wheaton & Lawson 1985). In refrigeration capacities less than about 350 kW the reciprocating compressor is slightly more efficient. The maximum refrigeration capacity of the largest reciprocating compressor available on the market in typical high stage application is approximately 900 kW (Stoecker 1998).

Centrifugal compressors are similar in design to a diffuser-type centrifugal water pump. Gaseous refrigerant is drawn into the centre of the rotating impeller. Centrifugal force imposed by the rotating impeller imparts high velocity to gas. As the gas passes through the diffusers the high velocity is converted to static pressure (Wheaton & Lawson 1985). Centrifugal compressors have been standard in large capacity chemical and process industry plants where they are driven by electric motors or by steam or gas turbines. Centrifugal compressors are also widely used for chilling water in air conditioning applications (Stoecker 1998).

The *condenser*: The condenser is a heat exchanger that accepts gaseous refrigerant at high pressure and temperature and removes enough heat to condense it to a liquid. The input and output condenser pressure and temperature very depending on the refrigerant used, system load and design of operating conditions (Wheaton & Lawson 1985). The refrigerant gives off heat in the condenser, and this heat is transferred to a medium having a lower temperature. The amount of heat given off is the heat absorbed by the refrigerant in the evaporator plus the heat created by compression input. The heat transfer medium can be air or water, the only requirement being that the temperature is lower than that which corresponds to the condensing pressure. The condenser may use air or water as the cooling medium. Air-cooled condensers are finned-tube radiators. Air flow is induced by fans. Water-cooled condensers are more compact, shell-and-tube type heat exchangers. The problems of fouling and scaling should be addressed by proper treatment of the cooling water in closed circuit (Danfoss 2007; Berk 2009).

Water-cooled condensers are usually of tube and shell type or the tube-in tube type. In shell and tube condensers cooling water is pumped through a serious of tubes that pass through a large container (shell). Refrigerant enters the shell and condenses on the tube exteriors. Because high water velocities are necessary in the tubes to get good heat transfer, contact time of water in the exchanger is short, even with counter flow conditions. Water temperature rise is thus small unless the contact time is increased. Shell and tube condensers have a good heat transfer characteristics as long as they are kept clean and water velocities of 1.3 to 2 m/s or higher are maintained. Tube-in-tube condensers consist of a large-diameter tube with a smaller-diameter tube centred inside. Cooling water flows between the two tubes. Tube length and diameters can be chosen to give any desired heat exchanger surface area and cooling capacity. By pumping water in one direction and refrigerant in the opposite direction, a counter current heat exchanger can be achieved (Wheaton & Lawson 1985). If a water-cooled condenser is used, the following criteria must be examined:

- requirement of cooling water for heat rejection
- utilization of a cooling tower if inexpensive cooling water is available
- requirement of auxiliary pumps and piping for recirculating cooling water
- requirement of water treatment in water recirculation systems
- space requirements
- maintenance and service situations, and
- provision of freeze protection substances and tools for winter operation (Dincer & Kanoglu 2010).

Evaporative condensers consist of a pipe through which the refrigerant passes. The pipe is usually formed into a grid to provide as much surface area per unit of space as possible. Evaporation of water, which is sprayed or dripped over the pipes, cools the refrigerant, condensing it from gas to a liquid. Cooling is by parallel current flow because the coolest water meets the warmest refrigerant. Cooling efficiency depends on several variables: including water flow rate, refrigerant flow rate, air circulation, air humidity, and temperature and heat transfer coefficient. Evaporative condensers are easily cleaned since all water-wetted surfaces are exposed, an important advantage if the cooling water has a high mineral content. Pumping costs are minimal and freezing is not a problem. The condenser size is larger than for other types with similar capacities (Wheaton & Lawson 1985). The following are some characteristics of these condensers:

- reduced circulating water for a given capacity
- water treatment is necessary
- reduced space requirement
- small pipe sizes and short overall lengths
- small system pumps, and
- availability of large-capacity units and indoor configurations.

The volume of water used by evaporative condensers is significant. Not only does water evaporate just to reject the heat, but water must be added to avoid the build-up of dissolved solids in the basins of the evaporative condensers. If these solids build up to the point that they foul the condenser surfaces, the performance of the unit can be greatly reduced (Dincer & Kanoglu 2010).

Air-cooled condensers are widely used, especially in smaller cooling systems. Their principal advantages are lighter weight and elimination of freeze-up problems. Cooling water is eliminated, a great advantage in areas in high mineral content water or areas with a water shortage. Air-cooled condensers are constructed of coils of tubing or tubing formed in grids. Most condensers have fins bonded to the tubes to extend their effective heat dissipation surface area. Fans are used to move air across the condenser tubes and improve cooling (Wheaton & Lawson 1985). The advantages of air-cooled condensers are the following:

- no water requirement
- standard outdoor installation
- elimination of freezing, scaling, and corrosion problems
- elimination of water piping, circulation pumps, and water treatment
- low installation cost
- low maintenance and service requirement
  high condensing temperatures
- high refrigerant cost because of long piping runs
- high power requirements per kW of cooling
- high noise intensity, and
- multiple units required for large-capacity systems (Dincer & Kanoglu 2010).

The *expansion valve*: The expansion valve is the heart of the refrigeration system. It meters the liquid refrigerant into the evaporator. Expansion valves may be fixed orifices, which cannot be adjusted after installation, or they can be automatically operated. Automatically operated expansion valves are controlled by temperature and pressure sensors usually located at or near the evaporator discharge. The expansion valve allows liquid refrigerant to flow into the evaporator where heat from the refrigerated area, is used to boil the liquid off as a gas. The temperature sensor is designed to hold the temperature at a value slightly above the refrigerant boiling point. The rise in temperature signals the expansion valve to open further allowing more liquid per unit time to flow into the evaporator. The expansion valve will open until the downstream evaporator temperature reaches the pre-set control temperature which is slightly above the refrigerant boiling temperature. The pressure sensor downstream of the evaporator corrects the pressure reading at the expansion valve owing to pressure drop across the evaporator. This is necessary since the boiling temperature of the refrigerant is a function of pressure (Wheaton & Lawson 1985). The most common throttling devices are

- thermostatic expansion valves
- constant-pressure expansion valves
- float valves, and
- capillary tubes.

The *thermostatic expansion valves* are essentially reducing valves between the high-pressure sides and the low-pressure side of the system. These valves, which are the most widely used devices, automatically control the liquid-refrigerant flow to the evaporator at a rate that matches the system capacity to the actual load. They operate by sensing the temperature of the superheated refrigerant vapour leaving the evaporator. For a given valve type and refrigerant, the associated orifice assembly is suitable for all versions of the valve body and in all evaporating temperature ranges (Dincer & Kanoglu 2010).

The *constant-pressure valve* is the forerunner of the thermostatic expansion valve. It is called an automatic expansion valve because of the fact that it opens and closes automatically without the aid of any external mechanical device. These expansion valves are basically pressure regulating devices. These valves maintain a constant pressure at outlet. They sense and keep the evaporated pressure at a constant value by controlling the liquid-refrigerant flow into the evaporator, based on the suction pressure. The refrigerant flows at a rate that exactly matches compressor capacity. Their applications are limited because of the constant cooling load (Dincer & Kanoglu 2010).

Float valves are divided into high-side float valves and low-side float valves. They are employed to control the refrigerant flow to a flooded-type liquid cooler. A high-side float valve is located on the high-pressure side of the throttling device. It is used in a refrigeration system with a single evaporator, compressor, and condenser. A low-side float valve is particularly located on the low-pressure side of the throttling device and may be used in refrigeration systems with multiple evaporators. In some cases, a float valve operates an electrical switch controlling a solenoid valve which periodically admits the liquid refrigerant to the evaporator, allowing the liquid level to fluctuate within preset limits (Dincer & Kanoglu 2010).

The *capillary tube* is the simplest type of refrigerant-flow-control device and may be used in place of an expansion valve. The capillary tubes are small-diameter tubes through which the refrigerant flows into the evaporator. These devices, which are widely used in small hermetic-type refrigeration systems (up to 30 kW capacity), reduce the condensing pressure to the evaporating pressure in a copper tube of small internal diameter (0.4–3 mm diameter and 1.5–5 m long), maintaining a constant evaporating pressure independently of the refrigeration load change. These tubes are used to transmit pressure from the sensing bulb of some temperature control device to the operating element. A capillary tube may also be constructed as a part of a heat exchanger, particularly in household refrigerators (Dincer & Kanoglu 2010).

The *evaporator*: The evaporator is the place where refrigeration is delivered to the system. Its geometry depends on the nature of the delivery. In a swept-surface freezer, it is the jacket of the heat exchanger. In a cold room, it is a finned-tube radiator type heat exchanger with fans. In a plate freezer, it is the hollow plates of the unit. In order to utilize fully the refrigeration capacity of the machine, it is important to secure that only gas (slightly superheated) leaves the evaporator (Berk 2009). Evaporators design varies with the application, but most designs fall into one of three general classifications: air systems, liquid systems and contact systems. Air systems are widely used in cold storage facilities, blast freezers and transport systems. Liquid systems include immersion freezing systems such as various brine chilling and freezing systems used on board fishing boats. Plate freezers are the most common contact systems used in the freezing of fish and the other aquatic products. Evaporators consist of a series of coils, either circular or more commonly in a rectangular pattern, which receive the low pressure liquid refrigerant from the expansion valve. Because of pressure low in the evaporator, the refrigerant boils and absorbs heat from the area around the evaporator coils. Air evaporator systems depend on the air to transfer heat from the product to evaporator. Achieving adequate heat transfer rates require the evaporator coils to operate 5–10°C below the desired air temperature. The temperature difference between air and coils create several problems in refrigeration or freezing of aquatic products. Quality maintenance is enhanced if air in the cooling or storage area approaches saturation. However, hot air holds more moisture than cooler air. The more moisture condensed on the coils the less efficient the refrigeration system, since refrigeration energy used to condense water does not cool the air. The lower humidity air also increases the risk of product dehydration, a condition often referred to as freezer burn. Evaporators using fans to force air across the coils are referred to as *forced air evaporators*. The fans may be either centrifugal or axial flow designs. Axial flow fans are probably most often used, except in areas where fan noise can be a problem. The fan forces air over the evaporator or by drawing the air over the coils in the opposite direction. The high air velocities over the evaporator coils increase heat transfer rates. Most evaporators using air as a heat transfer medium have fins attached to the evaporator coils. These fins are constructed of materials such as aluminium, which have a high conductivity. The fins increase the effective heat transfer area, and hence the heat transfer rate. Forced air systems are widely used in freezers, coolers and in both chilled and frozen storages. The major disadvantages of forced air systems are related to high air velocities within the cold room. *Natural convection evaporators* rely on the change in air density with temperature, cold air being heavier. Natural convection evaporators are generally located in the ceiling and in some instances on the walls. Warmer air rises to the evaporator where it is cooled. Once cooled it becomes heavy and falls causing a natural circulation. Natural convection evaporator require large surface areas since air velocities are low and heat transfer is retarded by the relatively thick air surface film on the evaporator surface. Systems using pipes or finned tubes as natural convection evaporators require considerable pipe or tubing length, which may take up a large volume in the cold room. *Liquid systems* differ from air systems in that het is transferred from product to evaporator by liquid rather than air. Liquids, usually brine, transfer heat much more efficiently than gases such as air. The heat transfer coefficient for liquids may be as much as ten times thus achievable with gases. Thus, evaporator surface area and physical size can be much smaller than for air systems while providing similar heat removal. Although brine systems are used only a limited amount for shore facilities, their primary use is onboard harvesting vessels to remove body heat from freshly caught fish. Contact systems used in aquatic products are almost all plate freezers of some type. Contact freezers depend on direct contact between modified evaporator coils and the product to transfer heat from product to refrigerant. The limiting factor in heat transfer with contact plate freezers is often heat conduction from the product interior to its surface (Wheaton & Lawson 1985).

Auxiliary hardware items, usually installed on the loop are: buffer storage vessels for refrigerant, refrigerant pumps for large distribution systems, oil and water separators, filters, internal heat exchangers, valves, measurement and control instruments, sight glasses etc. (Berk 2009).

6.3.2 Air-cycle system

The air-cycle system is different from the vapour-compression system in that the refrigerant (air) does not undergo the conversion from vapour to liquid and back again. In the air-cycle system, air from the refrigerated space enters a compressor, which increases the pressure of the air and temperature. The air then passes through a heat exchanger, which results in the high pressure air being cooled to a temperature near the outside ambient temperature. Next, the air moves through an expansion device that reduces the air pressure to that of the refrigerated space. The temperature of the air is also reduced by the expansion device to a value somewhat less than that of the refrigerated space. Finally, the chilled air is re-introduced to the refrigerated space, where it mixes with and lowers the temperature of the somewhat warmer air. Air continues to be drawn from the refrigerated space, passing through the refrigeration cycle, until the desired space temperature is achieved. A variation of the air-cycle system is used in commercial aircraft to provide cabin air conditioning. The compressed air is supplied by bleeding a small airstream from the compressor of the gas turbine engine providing propulsion. Cooling the high-pressure airstream is accomplished using a heat exchanger (coil) over which passes high-altitude ambient air. When the cooled high-pressure air is expanded to cabin pressure, the temperature decreases to somewhat below the cabin air temperature. The mixing of the expanded cool air with the cabin air cools the air in the cabin. To keep the cabin pressure from increasing, an equal amount of warm air is allowed to flow out as the chilled air is introduced. What differentiates the air-cycle system in aircraft cabin cooling is that the air does not complete a cycle, but rather begins as air entering the compressor of the engine and ends by leaking from the aircraft cabin. This type of cycle is usually referred to as an *open cycle* because the same air does not continue to circulate through the cycle (Fenton 2010).

6.3.3 Absorption system

An absorption system method of refrigeration is similar to a vapour-compression system in several respects. First, a refrigerant fluid sequentially moves through the components of the system. Second, the refrigerant moves through a condenser, expansion device, and evaporator in a manner similar to that of the vapour-compression cycle. The difference is in how the low-pressure, low-temperature refrigerant vapour is changed to a high-pressure, high-temperature vapour. In the absorption cycle, the vapour leaving the evaporator is absorbed by a liquid solution in a vessel called an absorber. Heat must be removed from the absorber to maintain the affinity that the vapour has for the solution. A pump increases the pressure of the liquid solution to the level of the condensing pressure; the liquid solution then enters another vessel called a generator. Heat is added to the generator, driving the refrigerant vapour out of the solution. At this point, the refrigerant enters the condenser, where the energy absorbed by the refrigerant is transferred to the atmosphere. The refrigerant then passes through an expansion device, which lowers its temperature and pressure. Finally, the low-temperature, low-pressure refrigerant enters the evaporator, completing the cycle. A significant quantity of heat at a moderate temperature is needed to operate the absorption system, and a relatively small input of electrical power is necessary to operate the pump. Thus, for example, when waste heat is available from a steam boiler, an absorption system may be employed to provide low-cost refrigeration. Water and ammonia are commonly paired in absorption refrigeration systems, with ammonia serving as the refrigerant. Because ammonia is the refrigerant, temperatures below the freezing point of water are possible. The success of the ammonia–water pair is due to ammonia's large affinity for water (Fenton 2010).

6.3.4 Thermoelectric system

In the 1960s, semiconductor materials were developed that allowed for commercial production of thermoelectric systems for refrigeration. These systems depend on the Peltier effect, first observed by Jean Peltier in 1834, which states that when an electric current passes through junctions of two dissimilar metals, one junction is cooled and the other is heated. Consequently, a cold surface is generated, whereupon heat may be absorbed. The hot surface receives the heat or thermal energy from the cold junction for transfer into the environment. Many junctions placed in series and attached to a plate forming a cold side and another plate for the hot side provide one approach to constructing a practical system (Fenton 2010).

6.3.5 Evaporative cooler

In climates where the air contains little moisture, the evaporative cooler is commonly used to cool residential and commercial buildings. The evaporative cooler typically consists of pads over which water is dripped. This promotes contact between the water and a warm dry air stream entering the evaporative cooler by means of a fan. The air is cooled by the evaporation of water. Reductions in air temperature may exceed −6.6°C (20°F), depending on how dry the outside air is. For evaporative cooling to be successful, the building must allow some conditioned air to exit, thereby allowing the cooled air to enter (Fenton 2010).

6.4 Refrigerants

A refrigerant is usually a substance that absorbs heat on expansion, and it should have the characteristics feature of changing from one phase to other on expansion or release of pressure. The refrigerants used in vapour compression cycles are volatile liquids. The desirable properties of a refrigerant are:

- *High latent heat of evaporation*: A relatively high evaporating temperature is required, so that heat transmission can occur with least possible circulating refrigerant.
- *High vapour density*: Refrigerant vapour ought not to have too high a specific volume because this is a determinant for compressor stroke at a particular cold yield.
- *Low pressure*: The refrigerant ought to have reasonable pressure, preferably a little higher than atmospheric pressure at the temperatures required to be held in the evaporator. The heavy refrigerator design

the pressure, which corresponds to normal condensing pressure, must not be too high.

- *Liquefied and evaporated at moderate positive pressure*: The refrigerant ought not to be corrosive and must not, either in liquid or vapour form, attack normal design materials.
- *Non-toxic, non-irritating, non-flammable, non-corrosive*: The refrigerant ought to not be poisonous. Where this is impossible, the refrigerant have a characteristic smell or must contain a tracer so the leakage can quickly be observed. The refrigerant ought not to be flammable or explosive. Where this condition cannot be met the same precautions as in the first point must be observed and local legislations must be followed.
- *Miscible with lubricating oils*: The refrigerant must not break down lubricating oil.
  - *Environmentally friendly.* The refrigerant must not pollute the atmosphere or other parts of the environment.
  - *Inexpensive, easily available*: The refrigerant must be easy to obtain and handle. The refrigerant must not cost too much.
  - *Stable*: The refrigerant must be chemically stable at the temperatures and pressures normal in a refrigeration plant (Berk 2009; Danfoss 2007; McGeorge 1995).

Needless to say no single fluid has all of these properties, and the choice of fluid any particular will always be a compromise (Hundy *et al.* 2008).

6.4.1 Classification of refrigerants

Refrigerants may be classified as either primary or secondary. A primary refrigerant cools the secondary loop in the primary–secondary heat exchanger. The secondary refrigerants circulated by a pump at relatively low pressure. Secondary refrigerants are generally liquids. For chilled application and frozen storage, brines of various types are usually used. Trichlorethylene, alcohols, glycols and halogen refrigerants are also used. Ammonia and halogen refrigerants are the most widely used primary refrigerants. Secondary refrigeration system reduces the amount of expensive primary refrigerant needed and reduces the line pressure in the cold storage room, a factor reducing the risk of leaks. Depending on the characteristics of the secondary refrigerant, leaks in the cold room may be less harmful. For example, ammonia a primary refrigerant is toxic to human and may damage the product if a leak develops in the cold room. The major disadvantages of secondary systems are the losses experienced in the primary–secondary heat exchanger and the need to operate the primary system at a lower temperature. The lower temperature is necessary because a 5 to 10°C temperature drop is necessary at both the primary–secondary heat exchanger and at the secondary heat exchanger in the cold room to maintain adequate heat transfer. Primary refrigerants should have a low boiling temperature and a high latent heat of vaporization at atmospheric pressure. Generally the higher latent heat of vaporization and the lower boiling point of a refrigerant the greater heat absorption per kilogram of refrigerant circulated through the system. In addition to good thermodynamic properties, a good refrigerant should be non-explosive, nontoxic, non-flammable, noncorrosive, non-irritating, non-injurious to foods, suitable for mechanical application, practically odourless, easily obtainable, inexpensive, efficient and economical use. A refrigerant must have a freezing point below the lowest temperature found in the system. The evaporating pressure in the evaporator should be above atmospheric pressure to prevent intrusion of air and moisture into the system. Low viscosity of both the liquid and vapour states is desirable to reduce friction losses and allow use of smaller pipes (Wheaton & Lawson 1985).

6.4.1.1 Halocarbons

The halocarbons contain one or more of the three halogens – chlorine, fluorine, or bromine – and are widely used in refrigeration and air-conditioning systems as refrigerants. These are more commonly known by their trade names, such as Freon, Arcton, Genetron, Isotron and Uron. Numerical indication is preferable in practice. In this group, the halocarbons were the most commonly used refrigerants (chlorofluorocarbons, CFCs). CFCs were commonly used as refrigerants, solvents and foam-blowing agents. The most common CFCs have been CFC-11 or R-11, CFC-12 or R-12, CFC-113 or R-113, CFC-114 or R-114, and CFC-115 or R-115. Although CFCs such as R-11, R-12, R-22, R-113 and R-114 were very common refrigerants in refrigeration and air-conditioning equipment, they were also used in several industries as aerosols, foams, solvents, etc. Their use rapidly decreased because of their environmental impact (Dincer & Kanoglu 2010).

6.4.1.2 Hydrocarbons (HCs)

HCs are the compounds that mainly consist of carbon and hydrogen. HCs include methane, thane, propane, cyclopropane, butane and cyclopentane. Although HCs are highly flammable, they may offer advantages as alternative refrigerants because they are inexpensive to produce and have zero ozone depletion potential (ODP), very low global warming potential (GWP), and low toxicity. There are several types of HC families, such as:

- Hydrobromofluorocarbons (HBFCs), compounds that consist of hydrogen, bromine, fluorine, and carbon.
- HCFCs are compounds that consist of hydrogen, chlorine, fluorine and carbon.
- Hydrofluorocarbons (HFCs) are compounds that consist of hydrogen, fluorine and carbon.
- Methyl bromide (CH_3Br) is a compound consisting of carbon, hydrogen and bromine.
- Methyl chloroform (CH_3CCl_3) is a compound consisting of carbon, hydrogen and chlorine (Dinçer & Kanoğlu 2010).

6.4.1.3 Inorganic compounds

Inorganic compounds are ammonia (NH_3), water (H_2O), air ($0.21O_2 + 0.78N_2 + 0.01Ar$), carbon dioxide ($CO_2$), and sulphur dioxide ($SO_2$). Among these compounds, ammonia has received the greatest attention for practical applications and, even today, is of interest.

Ammonia (R-717)

Ammonia, also referred to as refrigerant 717, has been used for many years as refrigerant because it has a high latent heat of vaporization and low boiling point at atmospheric pressure, and because it is inexpensive and readily available (Wheaton & Lawson 1985). Ammonia is used extensively in large industrial refrigeration plants. Its normal boiling point is −33°C. Ammonia has a characteristic smell, even in very small concentrations in air. It cannot burn, but it is moderately explosive when mixed with air in a volume percentage of 13 to 28% (Danfoss 2007). Because ammonia refrigeration systems use quite high pressure, the compressor and other equipment tend to be heavy, especially in smaller installations. Thus ammonia is used for most widely for larger refrigeration loads. Ammonia is neutral to iron and steel, which allows use of inexpensive pumping materials. In the presence of moisture, ammonia will corrode copper and brass alloys, precluding their use in ammonia systems. Ammonia is highly toxic to humans and may adversely affect food products. Leaks are easily detected because ammonia is readily detected by smell at very low concentrations. Ammonia also is not soluble in the most compressor oils, which may cause lubrication problems (Wheaton & Lawson 1985). Ammonia causes different technical and health problems. Gaseous ammonia is irritating to the eyes, throat, nasal passages, and skin. Ammonia reacts or produces explosive products with fluorine, chlorine, bromine, iodine and some other related chemical compounds. Ammonia reacts with acids and produces some heat. Despite its disadvantages, it is an excellent refrigerant and these possible disadvantages can be eliminated with proper design and control of the refrigeration system (Dincer & Kanoglu 2010).

Carbon dioxide (R-744)
Carbon dioxide is one of the oldest inorganic refrigerants. It is a colourless, odourless, nontoxic, non-flammable and non-explosive refrigerant and can be used in cascade refrigeration systems and in dry-ice production, as well as in food freezing applications (Dincer & Kanoglu 2010).

6.4.1.4 Refrigerant blends
Many of the HFC refrigerants are mixtures or blends of two or more individual chemicals. Mixtures can be azeotrope, near azeotropes or zeotropes. Azeotropes exhibit a single boiling point, strictly speaking at one particular pressure, but nevertheless they may be treated as a single substance. The first azeotropic refrigerant was a CFC, R502, so the use of refrigerant blends is not new. Where the boiling point varies throughout the constant pressure boiling process, varying evaporating and condensing temperatures exist in the phase change process (Trott & Welch 2000a; Hundy *et al.* 2008). Azeotropic blends are precise mixtures of substances that have properties differing from either of the two constituents. In a zeotropic mixture concentration of the two substances in the vapour is different from that in liquid at a given pressure and temperature (Stoecker 1998).

6.4.2 Ozone depletion potential

Some of the halocarbons are considered environmentally safe and others to be damaging. CFC is a non-hydrogenated halocarbon and it is extremely stable, which is a desirable feature for a refrigerant, but when released to the atmosphere it ultimately diffuses to the upper atmosphere. In the upper atmosphere it breaks down, the chlorine combines with ozone that exists there, depleting the ozone concentration. The second group of halocarbons is hydrogenated, because they contain a hydrogen atom. These are the HCFCs. Because of the hydrogen atom, the chemical is not quite as stable as a CFC, so when released to the atmosphere, most of it breaks down before reaching the ozone layer (Stoecker 1998).

The ozone layer in our upper atmosphere provides a filter for ultraviolet radiation, which can be harmful to human health. Researchers found that ozone layer was thinning, due to emissions into the atmosphere of CFCs, halons and bromides. The ODP of a refrigerant represents its effect on atmospheric ozone, and the reference point. Among the substances that deplete the ozone layer, refrigerant emissions are only about 10% of total; the remainder being made up of aerosol sprays, solvents and foam insulation. After agreement of the Montreal Protocol in 1987, the refrigeration industry rapidly moved from CFCs to HCFCs; R22 and HCFC replacement blends. With subsequent revisions of the protocol, a phase-out schedule for HCFCs was also set. R22, which is an HCFC, has a far lower ODP than the CFCs, but it was considered necessary to phase out all ozone depleting substances, and under the Protocol HCFCs will be eliminated by 2030. This signalled the end of R22 (Trott & Welch 2000a; Hundy *et al.* 2008).

6.4.3 Global warming potential

Global warming is possibly the most severe environmental issue faced by civilization today. The risk posed by its effects has been described in terms of environmental disaster due to huge future climate changes. Global warming is the increasing of the world's temperatures, which results in melting of the polar ice caps and rising sea levels. It is caused by the released into the atmosphere of so called 'greenhouse' gases, which form a blanket and reflect heat back to the surface of Earth, or hold it in the atmosphere. The refrigerant only affects global warming if released into the atmosphere. The choice of refrigerants affects the lifetime warming impact of a system and the term 'total equivalent warming impact' is used to describe the overall impact. It includes the effects of refrigerant leakage, refrigerant recovery losses and energy consumption (Trott & Welch 2000b; Hundy *et al.* 2008).

6.4.4 Safety of refrigerants

When dealing with any refrigerant, personal safety and safety of others are vitally important. Health and safety requirements are available from manufacturers of all refrigerants. HFC refrigerants are non-toxic in the traditional sense, but nevertheless great care must be taken to ensure adequate ventilation in areas where heavier than air gases may accumulate. Refrigerants are classified by toxicity and flammability hazard categories according to EN378, and safety codes are available from the Institute of Refrigeration for group A1 (low toxicity, non-flammable) groups A2/A3 (non-toxic and flammable), ammonia and carbon dioxide (Trott & Welch 2000b; Hundy *et al.* 2008).

6.5 Refrigeration of fish

Fish and other marine products are essentially composed mainly of proteins, fats and water, in which are dissolved many salts, sugars and other substances. When a fish dies bacteria begin to decompose the proteins, fats and other nutrients: the fats oxidize due to contact with air, and become rancid, and enzymes in the cells of fish and body fluids start to cause denaturation of the proteins. All these processes, which occur more quickly at higher temperatures, lead to fish spoilage and the purpose of refrigeration is to reduce the rate at which they occur (Preston & Vincent 1985).

The storage life of fresh perishable foods such as meat, fish, vegetables, and fruit can be extended by storing them at temperatures just above freezing, usually between 1 and 4°C for several days. Refrigeration slows down the chemical and biological processes in foods, and the accompanying deterioration and loss of quality and nutrients. Refrigeration also extends the shelf-life of products. The ordinary refrigeration of foods involves cooling only, without any phase change. Fish is a highly perishable commodity, and the preservation of fish starts on the vessel as soon as it is caught. Fish deteriorates quickly because of bacterial and enzymatic activities and refrigeration reduces these activities and delays spoilage. Different species have different refrigeration requirements, and thus different practices exist for different species. Large fish are usually eviscerated, washed and iced in the pens of the vessel's hold. Smaller fish, such as ocean perch and flounder, are iced directly without any processing. Lobsters and crabs are usually stored alive on the vessel without refrigeration. The fish caught in Arctic waters in winter are frozen by the cold weather and marketed as frozen fish. Salmon and halibut are usually stored in tanks of refrigerated sea water at 1°C. Fish are normally stored in chill rooms at 2°C as they await processing. They are packed in containers of 2- to 16-kg capacity in wet ice after processing and are transported to intended locations. It is desirable to keep fish uniformly chilled within 0 to 2°C during transit. At retail stores, fish should be displayed in special fish counters. Displaying in meat cases reduces the shelf-life of fish considerably, since the temperature of the meat case may be 4°C or higher. The maximum storage life of fresh fish is 10 to 15 days, depending on the particular species, if it is properly iced and stored in refrigerated rooms at 0 to 2°C. Temperatures below 0°C should be avoided in storage rooms since they slow ice melting, which can result in high fish temperatures. Also, the humidity of the storage facility should be over 90% and the air velocity should be low to minimize dehydration (Stoecker 1998). The spoilage of fish due to protein denaturation, fat changes and dehydration can all be slowed down by reducing the storage temperature.

6.6 Refrigeration on board

Fish begin to spoil immediately after death. Efficient methods of preservation on board fishing vessels are necessary in order to land fish of good quality and permit long voyages. Since the rate of spoilage is largely dependent on temperature, increased by increase in temperature, refrigeration of the catch is common practice (Merritt 1969).

Most fish is caught at sea and must be cooled soon after it is taken on board, and kept cold until it can be sold, frozen or otherwise processed. The general practice is to put the fish into refrigerated sea water tanks, kept down to 0°C by direct expansion coils or a remote shell-and-tube evaporator. The sea water must be clean and may be chlorine dosed. At this condition, fish can be kept for up to four days. Ice is also used on board, carried as blocks and crushed when required, carried as flake, or from shipboard flake-ice makers. Fresh fish is stored and transported with layers of ice between and over the fish, cooling by conduction and keeping the product moist. Fish kept at chill temperatures in this manner can travel to the final point of sale, depending on the time of the journey. Where refrigerated storage is used, the humidity within the room must be kept high, by using large evaporators, so that the surface of the fish does not dry (Trott & Welch 2000b).

Lowering of the body temperature as soon as possible is most important to preserve the initial freshness of fish. All fish should be chilled rapidly and immediately after catch. Stowage in ice is the most widely used method of chilling either in box on shelves or in ponds inside fish-hold. When large quantities of fish are caught they are difficult to handle quickly with normal icing methods, so chilled seawater (CSW) or refrigerated seawater (RSW) systems are used. It is important to be able to calculate the amount of ice needed for chilling and storing a given amount of fish. In practice, however, there are many factors that affect this calculation such as: fish hold's wastage of ice stock before it can be used; insulation of the fish hold; warming during loading and unloading, all of which makes it more difficult to estimate the actual amount of ice required. Ice should be used economically and effectively. On fishing boats, the insulation properties of the fish hold are most important. It is often noted that a new boat will keep the fish fresh and for longer, but after several voyages it may become poor and less efficient due to the ageing of the insulation system and bacterial contamination of the fish hold. Some fishing boats are also equipped with a refrigeration machine. This does prevent the ice from melting, but care must be taken not to lower the storage room temperature below freezing point. An advantage of crushed-ice-storage is that the ice does not only cool the fish body, but also the thawing water runs down the surface of the fish washing away the bacteria in the slime, and protects the body surface from the air, which may cause dehydration, rancidity and oxidation. If, however, the cooling machine lowers the room temperature to as low as −30 to −40°C, the crushed ice will not melt and the fish muscle become frozen, which will injure the texture. When the cooling machine is running, more care must be taken if thawed water is constantly being discharged as the fish around the cooling pipes are apt to get frozen. When storing a large catch in the fish hold, some boards should be laid at intervals between the fish so that the fish on the bottom will not be crushed (SEAFDEC 2005).

Whatever the refrigeration system, its function is to reduce the fish temperature as necessary, usually quickly, and then maintain the required temperature against the ingress of heat. In order to reduce the temperature, heat must be removed. Icing is a common method to reduce fish temperature onboard application. Onboard icing was explained in Chapter 5. The fish boxed with ice should be maintained under refrigerated conditions in order to prevent bacterial and enzymatic changes. A low fish room temperature will help prevent spoilage of fish. Even with proper icing techniques, however, the introduction of mechanical refrigeration, particularly in a warm climate, can enable a reduction in the amount of ice required. Thus the work of stowage may be made easier and, where the size of fish room places a limitation on the duration of voyage, the amount of fish landed may be increased. The fact that mechanical refrigeration can eliminate variations in icing technique from one season to another also will be an advantage in itself, notably where the fish are boxed at sea. Use of mechanical refrigeration as a supplement to melting ice should be considered carefully in the light of climate, size of vessel, type of fishery, market, length of voyage, cost of ice, and possibly some other factors (Merritt 1969).

6.6.1 Refrigeration capacity

Refrigeration plants are used onboard mainly for food preservation in cold rooms and for producing ice required for onboard consumption. Heat load calculations provide the refrigeration capacity required for cold rooms. Heat load can be calculated by using methods available in scientific literature of commercial software. The required capacity of a refrigeration plant is determined from individual cold room heat load, ice production load and from the required cooling time period (Amey & Majgaonkar 2008).

The refrigeration system capacity is justified by energy of heat which unit in BTU or kilocalories (kcal) in a specified period of time like (per day or per hour). In general, the refrigeration capacity is expressed in units of kcal/h, British thermal units (BTU)/h or tonne.

A common standard unit is express for 1 tonne of refrigerant is represent for a capacity of heat 12 000 BTU (SEAFDEC 2005).

Tuna quality is preserved only by the proper sequence of chilling and freezing on board the vessel. Preservation of quality must starts at point of capture because quality loss begins when fish dies and it can never be improved, but it can be maintained with proper care. Tuna are commonly chilled and stored temporarily in RSW, however, rapid chilling and continued freezing protects tuna quality. Spoilage and quality changes of fish are affected by temperature, time and physical treatment. High temperature increases the rate of spoilage, while low temperatures slow spoilage. The rapid and careful handling, from the time the fish are in the net until frozen in the well, will help determine their final quality. The safest way to ensure that quality is maintained is to match the vessel's capture rate to its refrigeration capacity, which is its rate or capacity to freeze the fish as quickly as possible. A vessel's capacity is not the unfilled wells but rather the available refrigeration to preserve the fish coming aboard in the manner dictated by its condition. When fishing is heavy, the catch rates of good quality fish being brought onboard needs to be matched with the vessel's refrigeration capacity and the engineer's abilities to handle and freeze at any point in time (Anonymous 2005). The mechanical refrigeration system aboard tuna seiners was developed in the late 1930s for use aboard bait boats and some vessels, less than 30 m size, had been equipped with onboard mechanical refrigeration and freezing in the 1960s. Since the refrigeration system in use aboard tuna seiners was developed in the late 1930s, fishing methods have changed, vessel carrying capacities have increased tenfold, catch rates have increased fivefold, and individual fish well sizes have increased fourfold The well capacities are thus in the order of 44–94 short tonnes of fish per well. Each well is lined on the sides and overhead with evaporator coils, and each well is insulated with polyurethane foam. The refrigeration system aboard these vessels is a direct expansion ammonia system powered by reciprocating compressors with condensation provided by vertical-tube, box-type condensers. Receivers are provided below the condensers to contain reserve ammonia. This system uses circulated seawater and brine as the secondary refrigerant to chill and freeze the tuna. The existing system is a simple single-stage ammonia system with seawater or brine acting as the secondary refrigerant. Analysis of the present systems aboard tuna seiners indicates that the ammonia compressor capability is adequate, but that the evaporation ability is the limiting factor in providing adequate refrigeration in the larger fish wells. The principal difference between the sorption and the vapour-compression cycles is the mechanism for circulating the refrigerant through the system and providing the necessary pressure difference between the vaporizing and condensing processes. The vapour-compressor employed in the vapour-compression cycle is replaced in the sorption cycle by a sorber and a generator, which compress the vapour as required. However, it is difficult to use an absorption system on a smaller fishing boat because absorption systems have complex components, large volume, high initial costs and are inconvenient to control. In particular, all four main components in the system (generator, condenser, absorber and evaporator) have a free liquid level, and are thus unsuitable for the reeling of the hull. Refrigeration equipment for a fishing boat is required to be as simple as possible and very reliable. On these points the adsorption system is better than any other (Wang & Wang 2005).

6.7 Combination of refrigeration with traditional and advanced preserving technologies

Combination of treatments for food preservation may result in synergistic or cumulative effects of microbiological barriers or hurdles, leading to a reduced level of one or all the treatments (Leistner & Gorris 1995).

High hydrostatic pressure (HHP): HHP treatment is of great value to the seafood industry processing and has been applied to a range of different seafood; HHP technology offers food processors several advantages over conventional processing methods. For example, pressure is transmitted instantaneously and uniformly throughout a system. HHP processing has the potential to increase their shelf-life (Murchie *et al.* 2005). If the product preserved by HHP was stored at low temperature, its shelf-life and quality will be improved. It was reported that high hydrostatic pressure treatment was much less effective at 12°C compared to 3.5°C in pike (Krizek *et al.* 2014) and trout (Matejkova *et al.* 2013). The use of high pressure in combination with low temperature on growth and proliferation of Listeria in smoked salmon mince has been reported (Picart *et al.* 2005).

Modified atmosphere packaging (MAP): Temperature is of primary importance in all fresh fish storage, including MAP and vacuum packaging, as both enzymatic and microbiological activity are greatly influenced by temperature. Many bacteria are unable to grow at temperatures below 10°C and even psychotropic organisms grow very slowly, and sometimes with extended lag phases, as temperatures approach 0°C. Increased solubility of CO_2 at lower temperatures will relatively increase the effect of MAP (Sivertsvik *et al.* 2002). MAP and vacuum packaging, with refrigeration, have received increasing attention as methods of food preservation (Masniyom 2011; Masniyom *et al.* 2013). Shelf-life of iced or refrigerated fish could be extended by using MAP, specifically elevated CO_2 levels, which has shown to retard the growth of spoilage and pathogenic bacteria. Vacuum packaging is used for long-term storage of dry foods and the shelf-life extension of seafood. The product is packed in a vacuum package which has good barrier properties towards oxygen and water and is easily sealed. Air is removed under vacuum and the package is sealed. Furthermore, vacuum packaging could prevent oxidative rancidity and improve organoleptic quality in seafood. The shelf-life of fresh fish can be doubled or tripled by treating the fish with ionizing radiation. Radiation has no adverse effect on quality, but consumer resistance has prevented its widespread use. The shelf-life can also be extended by inhibiting the growth of bacteria by storing the fish in a modified atmosphere with high levels of carbon dioxide and low levels of oxygen (Masniyom *et al.* 2013). Longer shelf-life at 0°C for haddock packed under MAP (40 or 60% CO_2) compared with air at 5 and 10°C has been reported (Dhananjaya & Stroud 1994). Arkoudelos *et al.* (2007) reported that MAP mixture of CO_2 (40%), N_2 (30%) and O_2 (30%) in combination with refrigeration (0°C) was the most effective treatment for the preservation and in extending the shelf-life of raw eel (*Anguilla anguilla*), followed by vacuum packaging and refrigeration. *Pangasius hypophthalmus* fillets were packed under two modified atmosphere conditions (50% CO_2, 50% N_2 and 50% CO_2, 50% O_2) and it was found that the use of CO_2 in modified atmosphere significantly prolonged the shelf-life of fillets stored under refrigeration (Noseda *et al.* 2012).

Irradiation: Food irradiation, in combination with good refrigeration and handling practices, might provide a means to increase fish product shelf-life. Food irradiation is a process that has proven to be successful, not only in ensuring the safety, but also in extending the shelf-life of fresh meats because of its high effectiveness in inactivating pathogens without deterioration in product quality (Arvanitoyannis *et al.* 2009). The preservative effects of ionizing radiation can often be combined advantageously with the effects of other physical and chemical agents. The resulting combination treatments may involve synergistic or cumulative action of the combination partners, leading to a decreased treatment requirement for one or both the agents. This, in turn, may result in savings in both cost and energy and may bring about an improvement in the sensory properties and bacteriological quality of the food thus treated. Preservative effects of combinations of treatments in controlling microbial growth and resulting spoilage is based on hurdle technology and involves the creation of series of hurdles in the foods for microbial growth. Such hurdles include heat, irradiation, low temperature, water activity, and pH, redox potential and chemical preservatives. It was reported that irradiation, with a combination of low activity and low temperature, could ensure the safety of kwamegi, (semi-dried raw Pacific saury) (Chawla *et al.* 2003). Combination of irradiation and refrigeration in carp (*Cyprinus carpio*) showed synergic affect to prolong the shelf-life (Icekson *et al.* 1996).

Biopreservation: Biopreservation is a technology used to extend the shelf-life and/or control the growth of pathogenic flora of refrigerated products by the inoculation of bacteria selected for their inhibition properties towards undesirable bacteria (Leroi *et al.* 2006). Lactic acid bacteria (LAB) are usually chosen for these applications as they produce a wide range of inhibitory compounds such as organic acids, hydrogen peroxide, diacetyl and bacteriocins, or compete with other micro-organisms by nutrient depletion (Leroi 2011).

Coating of film: Edible coatings can be used to prolong the shelf-life of refrigerated fish. Chitosan-based coatings have already been used for a variety of fish species to reduce microorganisms improve overall fish quality and extent its shelf-life. Chitosan can also be combined with various functional compounds that have antioxidant, antimicrobial or other activities to enhance the product quality. Several studies have been conducted on use chitosan coating of fish and fish products to extent shelf-life under refrigeration storage. Significant increase in the shelf-life of Japanese sea bass (*Lateolabrax japonicas*) treated with chitosan incorporated with citric acid and licorice extract (Qiu *et al.* 2014). Chitosan was effective in reducing lipid oxidation and microbial growth in herring and Atlantic cod (Jeon *et al.* 2002) and provided partial protection by delaying spoilage of fish patties stored at 2°C (Lopez-Caballero *et al.* 2005). Dipping with tea polyphenol and rosemary extract combined with chitosan coating were effective for the preservation of large yellow croaker during refrigerated storage, and could extend the shelf-life of fish by 8–10 days compared with the control group (Li *et al.* 2012). Shelf-life of hake and sole was extended by coating chitosan under refrigeration (Fernandez-Saiz *et al.* 2013).

References

Amey, S. & Majgaonkar, M.E. (2008). Refrigeration for ships. *Ashrae Journal*, December, 50–51.

Anonymous (2005). Recommendations for On-board Handling of Purse Seine-Caught Tuna. UTSF Technical Bulletin. LMR Fisheries Research, Inc., San Diego, CA.

Arkoudelos, J., Stamatis, N. & Samaras, F. (2007). Quality attributes of farmed eel (*Anguilla anguilla*) stored under air, vacuum and modified atmosphere packaging at 0°C. *Food Microbiology*, 24: 728–735.

Arvanitoyannis, I.S., Stratakos, A. & Mente, E. (2009). Impact of irradiation on fish and seafood shelf life: A comprehensive review of applications and irradiation detection. *Critical Reviews in Food Science and Nutrition*, 49: 68–112.

Berk, Z. (2009). Refrigeration, equipments and methods. In: *Food Process Engineering and Technology* (Ed Z. Berk), 1st edn, pp. 413–426, Elsevier Inc., New York.

Chawla, S.P., Kim, D.H., Jo, C., Lee, J.W., Song, H.P. & Byun, M.W. (2003). Effect of gamma irradiation on the survival of pathogens in Kwamegi, a traditional Korean semidried seafood. *Journal of Food Protection*, 66 (11): 2093–2096.

Danfoss (2007). Refrigeration: an introduction to basics. Refrigeration and Air Conditioning Division. Lecture. Nordborg, Denmark.

Dincer, I. & Kanoglu, M. (2010). Refrigerants. In: *Refrigeration Systems and Applications* (Eds Dincer, I. & Kanoglu, M.), 2nd edn, pp. 63–105, John Wiley & Sons, Ltd, Chichester, UK.

Dhananjaya, S. & Stroud, G.D. (1994). Chemical and sensory changes in haddock and herring stored under modified atmosphere. *International Journal of Food Science and Technology*, 29: 575–583.

Fenton, D.L. (2010). *Fundamentals of Refrigeration* (Ed. Fenton, D.L.), 2nd edn, pp. 1–371, American Society of Heating, Refrigerating and Air-Conditioning Engineers, Inc. (ASHRAE), NE, Atlanta GA.

Fernández-Saiz, P., Sánchez, G., Soler, C., Lagarona, J.M. & Ocioa, M.J. (2013). Chitosan films for the microbiological preservation of refrigerated sole and hake fillets. *Food Control*, **34**: 61–68.

Hundy, G.H., Trott, A.R. & Welch, T.C. (2008). Refrigerants. In: *Refrigeration and Air Conditioning*. 4th edn, pp. 30–41, McGraw-Hill, Oxford.

Icekson, I., Pasteur, R., Drabkin, V., Lapidot, M., Eizenberg, E., Klinger, I. & Gelman, A. (1996). Prolonging shelf-life of carp by combined ionising radiation and refrigeration. *Journal of the Science of Food and Agriculture*, **72** (3): 353–358.

Jeon, Y.J., Janak, Y.V., Kamil, A. & Shahidi, F. (2002). Chitosan as an edible invisible film for quality preservation of herring and Atlantic Cod. *Journal of Agricultural and Food Chemistry*, **50**: 50167–5178.

Krizek, M., Matejkova, K., Vacha, F. & Dadakova, E. (2014). Biogenic amines formation in high-pressure processed pike flesh *(Esox lucius)* during storage. *Food Chemistry*, **151**: 466–471.

Leistner, L. & Gorris, L.G.M. (1995). Food preservation by hurdle technology. *Trends in Food Science and Technology*, **6**: 41–46.

Leroi , F., Amarita, F., Arboleya, J.C., *et al.* (2006). Hurdle technology to ensure the safety of seafood products. In: *Improving Seafood Products for the Consumer* (Ed. Børresen, T.), pp. 399–425, Woodhead Publishing Limited, Cambridge, UK.

Leroi, F. (2011). Biopreservation of lightly preserved seafood Products. *Infofish International*, **4**: 41–46.

Li, T., Hub, W., Li, J., Zhanga, X., Zhua, J. & Li, X. (2012). Coating effects of tea polyphenol and rosemary extract combined with chitosan on the storage quality of large yellow croaker (*Pseudosciaena crocea*). *Food Control*, 105–101–106.

Lopez-Caballero, M.E., Gomez-Guillen, M.C., Perez-Mateos, M. & Montero, P. (2005). A chitosan–gelatin blend as a coating for fish patties. *Food Hydrocolloids*, **19**: 303–311.

Masniyom, P. (2011). Deterioration and shelf-life extension of fish and fishery products by modified atmosphere packaging. *Songklanakarin Journal of Science and Technology*, **33**: 181–192.

Masniyom, P., Benjama, O. & Maneesri, J. (2013). Effect of modified atmosphere and vacuum packaging on quality changes of refrigerated tilapia (*Oreochromis niloticus*) fillets. *International Food Research Journal*, **20** (3): 1401–1408.

Matejkova, K., Krizeka, M., Vácha, F. & Dadakovaa, E. (2013). Effect of high-pressure treatment on biogenic amines formation in vacuum-packed trout flesh *(Oncorhynchus mykiss)*. *Food Chemistry*, **137**: 31–36.

McGeorge, H.D. (1995). Refrigeration. In: *Marine Auxiliary Machinery*, 7th edn, pp. 333–367, Butterworth-Heinemann, Oxford.

Merritt, J.H. (1969). Mechanical refrigeration with ice. In: *Refrigeration of Fishing Vessels*, pp. 57–77, Fishing News Books ltd., London, UK.

Murchie, L.W., Cruz-Romero, M., Kerry, J.P., *et al.* (2005). High pressure processing of shellfish: A review of microbiological and other quality aspects. *Innovative Food Science and Emerging Technologies*, **6**: 257–270.

Noseda, B., Islam, M.T., Eriksson, M., *et al.* (2012). Microbiological spoilage of vacuum and modified atmosphere packaged Vietnamese *Pangasius hypophthalmus* fillets. *Food Microbiology*, **30**: 408–419.

Picart, L., Dumay, E., Guiraud, J.P., & Cheftel, J.C. (2005). Combined high pressure-sub-zero temperature processing of smoked salmon mince: Phase transition phenomena and inactivation of *Listeria innocua*. *Journal of Food Engineering*, **68**: 43–56.

Preston, G.L. & Vincent, M.A. (1985). Fisheries sector refrigeration systems in Pacific Island countries. Seventeenth Regional Technical Meeting of Fisheries 5–9 August. South Pacific Commission, Noumea, New Caledonia.

Rao, D.G. (2010). Refrigaration. In: *Fundamentals of Food Engineering* (Ed. Rao, D.G.), pp. 52–64, PHI Learning Private Limited, New Delhi, India.

Qiu, X., Chen, S., Liu, G. & Yang, Q. (2014). Quality enhancement in the Japanese sea bass (*Lateolabrax japonicas*) fillets stored at 4°C by chitosan coating incorporated with citric acidor licorice extract. *Food Chemistry*, 162–156–160.

SEAFDEC (2005). Chilling systems by refrigerated sea water. In: Onboard Fish Handling and Preservation Technology, pp. 6–9. Southeast Asian Fisheries Development Center, Samut-Prakan, Thailand.

Sivertsvik, M., Jeksrud, W.K. & Rosnes, J.T. (2002). A review of modified atmosphere packaging of fish and fishery products-significance of microbial growth, activities and safety. *International Journal of Food Science and Technology*, **37**: 107–127.

Stoecker, W.F. (1998). Reciprocating compressors. In: *Industrial Refrigeration Handbook*, pp. 93–121, McGraw-Hill Education, New York.

Stoecker, W.F. (1998). Refrigerants. In: *Industrial Refrigeration Handbook*, pp. 415–434, McGraw-Hill Education, New York.

Stringer, M. & Dennis, C. (2000). Refrigeration of c hilled foods. In: *Chilled Foods: A Coprehensive Guide* (Eds Stringer, M. & Dennis, C.), 2nd edn, pp. 79–90, Woodhead Publishing Limited and CRC Press LLC, Cambridge, UK.

Trott, A.R. & Welch, T.C. (2000a). Refrigerants. In: *Refrigeration and Air Conditioning*, 3rd edn, pp. 28–35, Butterworth-Heinneman, Oxford.

Trott, A.R. & Welch, T.C. (2000b). Refrigeration in the food trades-meats and fish. In: *Refrigeration and Air Conditioning*, 3rd edn, pp. 188–192, Butterworth-Heinneman, Oxford.

Wang, S.G. & Wang, R.Z. (2005). Recent developments of refrigeration technology in fishing vessels. *Renewable Energy*, **30**: 589–600.

Warren, P. (1986). The chilled fish chain. An Open Learning Module for The Sea fish Open Technical Project. Her Majesty's Stationery Office, London.

Wheaton, F.W. & Lawson, T.B. (1985). Refrigerated process. In: *Processing Aquatic Food Products* (Eds Wheaton, F.W. & Lawson, T.B.), pp. 181–222, John Wiley & Sons Inc., New York.

CHAPTER 7

Freezing technology

7.1 Principles of freezing

Freezing is the process that lowers the temperature to below the freezing point. These temperatures allows most of the water to turn into ice. The freezing point depends on the substances dissolved in the fluid of the tissue. Fish contain 75–80% water. The freezing process concerns the removal of latent heat during the phase transition of water from liquid to solid and removal of sensible heat, depending on the reduction of temperature.

Dissolved and colloidal materials in the fluid of the tissue decrease the freezing point to below 0°C. The tissue fluid is a solvent containing various dissolved materials. Therefore, the water in food starts freezing at a certain temperature, and passes through many cryohydric points, depending on the solutes and freezes at the lowest cryohydric temperature. Each food needs to be cooled down to the last specific cryohydric point in order to freeze completely. Seafood starts freezing at the temperatures between –1°C and –3°C. Water turns into ice during the freezing process and the concentration of organic and inorganic salts depresses the freezing point. Most of the water (90–95%) is frozen at –25°C, but much of it is turned into ice between –1°C and –5°C; this

Seafood Chilling, Refrigeration and Freezing: Science and Technology, First Edition.
Nalan Gökoğlu and Pınar Yerlikaya.
© 2015 John Wiley & Sons, Ltd. Published 2015 by John Wiley & Sons, Ltd.

is called the freezing zone or critical zone. This zone should be passed quickly to obtain small ice-crystals so that formation of large ice-crystals and their disruptive effects on cell walls are avoided. The decrease in product temperature changes with time, because of the removal of the latent heat of ice-crystal at −1/−5°C and increasing thermal diffusivity of the food, due to ice-crystals. The increase in ice formation is slower at temperatures below −10°C.

The freezing process of fish takes place in three stages. First, the temperature of the product decreases just below 0°C. Cooling to the freezing point removes the sensible heat. Second, temperature remains constant at around −1°C, which is defined as the 'thermal arrest period'. Latent heat is removed. The fish should pass through the thermal arrest period as soon as possible in order to produce a good-quality frozen product. In the last stage, the temperature of the product decreases sharply, allowing freezing of the remaining water. This further cooling to the desired subfreezing temperature removes the sensible heat of frozen food (Gökoğlu 2002; Venugopal 2006).

7.1.1 Water and ice

Water is a direct reactant in metabolic processes and supports chemical reactions. The basic reason for the rapid deterioration of marine-derived foods is their water content of up to 95%. This phenomenon allows microorganisms to utilize water because water needs to be in liquid form in order to be used by microorganisms. In the freezing process, the food material becomes unfavorable for the utilization of water due to decreasing water activity by turning into ice. Water activity is a measure of the degree to which water is bound within the food, and thus unavailable for further chemical or microbiological activity. The relationship between water activity and moisture content for foods at a particular temperature is a sigmoidal-shaped curve called the sorption isotherm (Figure 7.1). The water activity scale is 0.00 to 1.00, where 1 is pure water and completely available for microorganisms (Abbas *et al.* 2009; Cybulska & Doe 2002).

Water in food divided into tightly bound, loosely bound and non-bound waters. Decreased water activity retards the growth of microorganisms, slows enzyme-catalysed reactions and retards non-enzymatic browning (Belitz *et al.* 2009). The freezing process allows water to turn into ice and inhibits its usage as a reactant or solvent in chemical and biochemical reactions.

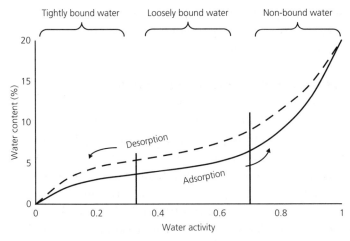

Figure 7.1 Moisture sorption isotherm curve of a food product.

7.1.2 Nucleation in pure water

Energy is removed during the transition from liquid water to ice in the freezing process. The free substitution of molecules in the liquid phase decelerates, and tends to accelerate spontaneously in order to organize a regular structure. This is the primary signs of phase transition. Phase change of a pure material depends on a specific temperature of its own. However, reaching that the freezing temperature of the liquid does not initiate the phase transition; structures named 'nuclei' should be in the medium for the start of the phase change. If the nuclei are not available in the medium, the liquid should be cooled below to the critical temperature of its specific freezing point. Small clusters of water molecules form the basis for crystal-initiating nuclei, which is known as homogeneous nucleation. The rate of homogeneous nucleation increases with the degree of supercooling. If the driving force for nucleation is an impurity then this is called heterogeneous nucleation.

7.1.3 Freezing point depression

Ice formed from a dilute solution does not contain the solute. Thus, the vapour pressure of ice at various temperatures does not change, but the vapour pressure of a solution is lower than that of ice at the freezing point. Further cooling is required for ice to form, and the net lowering of vapour pressure is called freezing point depression (Chieh 2006). This phenomenon can be confused with supercooling. Supercooling is the cooling of a liquid below its freezing point, without it becoming solid (see next section). Freezing point depression is when the solution is cooled below the freezing point of the corresponding pure liquid due to the presence of the solute.

7.1.4 Crystallization and crystal growth

The nucleation of water occurs in an unstable manner. Small-sized crystals have no chance to grow and leave their molecules into the liquid phase and disappear. The crystals reaching to the critical size maintains and/or reaches to larger dimensions. During this thermodynamic unstable state, the temperature drops below to 0°C without phase change; this is called 'supercooling' (the difference between the ambient temperature and that of the solid–liquid equilibrium) (Alizadeh *et al.* 2007). The reason for supercooling is the delay in nucleation. Supercooling is the driving force of the nucleation, and is an important parameter that controls the size and number of ice-crystals (Figure 7.2).

The nucleation starts when the supercooled water reaches a specific temperature and is followed by the expansion of crystals. Release of the latent heat of freezing allows the temperature to increase 0°C, and should be removed in order to perform an exact freezing. This process accelerates the freezing rate.

The rate of crystal growth is affected by the temperature of the medium, the removal of heat, and the temperature difference between ice-crystal surfaces and the cooling medium. The nucleation sites

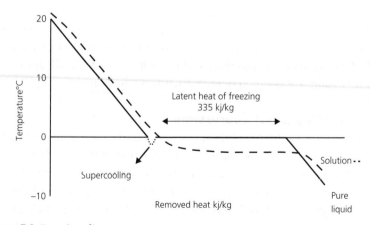

Figure 7.2 Freezing diagram.

provide the nuclei for growth of large crystals at low cooling rates, whereas numerous nucleation sites are provided with rapid freezing. Rapid cooling results in small ice-crystals and avoids mechanical disruption of cell and tissues. It also minimizes transport of water from inside the cells to the surrounding extracellular spaces.

The growth rate of crystals and their mass are different phenomena. The mass of crystals depends on the initial number of nuclei. A greater number of nuclei allows the ice-crystals to remain small. If the temperature of the medium decreases rapidly and the latent heat of freezing can be removed in a short time, numerous small crystals are formed. Ice-crystal size depends on the rate of freezing; the faster the rate the more nucleation is promoted (Gökoğlu 2002; Karel & Lund 2003; Erkan & Varlık 2004; Cemeroğlu & Soyer 2005). The size of the crystals varies depending on the product, likewise pre-rigor and post-rigor fish when frozen at the same rate, will exhibit a difference in the mean size of crystals.

7.1.5 Recrystallization

The sizes of the crystals are not stable and always tend to grow. The changes in the number, size, shape and orientation of the crystals during storage after freezing are defined as recrystallization. The driving force for recrystallization is the tendency of all systems to minimize their free energy. This tendency is manifested primarily by the increase in crystal size and widening of crystal size distribution.

Recrystallization basically involves small crystals disappearing, large crystals growing, and crystals fusing together;

Oswald ripening, or migratory recrystallization, refers in general to the tendency of larger crystals to grow at the expense of smaller crystals as a result of the difference in their melting points. Larger particles are more energetically favoured than smaller particles. Fluctuating temperatures greatly enhance the process of recrystallization. The equilibrium between ice and liquid form is overbalanced to the liquidus line in case of temperature fluctuation. Increasing temperature allows the melting of ice-crystals. In the reverse process, nucleation of new crystals is formed and equilibrium is re-established by crystal growth. Melt–refreeze behaviour occurs to a greater extent at higher temperatures and leads to complete disappearance of smaller crystals. As a result, the number of ice-crystals decreases and their mean size increases with time.

Iso-mass recrystallization (rounding off) means that the crystals with irregular shapes and large surface-to-volume ratios tend to become a more compact spherical shape since a sphere is the geometric shape with minimum surface to volume ratio.

Fusion or acceleration of ice-crystals is another recrystallization mechanism. Two growing crystals come together and form into a single ice-crystal.

Recrystallization, regardless of type, can be minimized by maintaining a low and constant storage temperature. Recrystallization may take place even during thawing. The thawing process should be done rapidly to avoid the undesirable effects of recrystallization.

7.1.6 Freezing time

It is important to know the freezing time regardless of the freezing method in terms of energy consumption, efficient utilization of cold stores, product quality and economic evaluations. Freezing time depends on the nature of the fish including thermal conductivity, thermal resistance and its shape. The enthalpy of fish is related to the water fraction and temperature. For thermal calculations such as freezing and thawing, the differences in enthalpy are essential (Keizer 1995).

The freezing time is the time taken to lower the temperature of the product from its initial temperature to a given temperature at its thermal centre. It is not easy to calculate the freezing time because of the alteration of thermal aspects of the material during freezing process. Also, the food is not homogeneous and mostly do not have a regular shape. In general, the Planck equation is used to estimate the freezing time of a product.

$$t_f = \frac{\rho H_L}{T_f - T_\infty} \times \left(\frac{PL}{h_c} + \frac{RL^2}{\lambda_d} \right)$$

where: t_f is freezing time (s); ρ is specific volume of fish (kg/m); H_L is the latent heat of freezing (J/kg); T_f is the initial freezing point of the fish (°C); T_∞ is the refrigeration medium temperature (°C); L is the thickness of the product in direction of prevailing heat transfer (m); h_c is the surface heat transfer coefficient (W/m/°C); λ_d is the coefficient of thermal conductivity of frozen fish (W/m/°C); P and R are constants depending on the shape of the material (P sphere, infinite cylinder, infinite slab −0.167, 0.167, 0.500, respectively; R sphere, infinite cylinder, infinite slab −0.042, 0.042, 0.250, respectively).

The Planck equation is simple and clear. The presence of wrappings, and many other factors, should be considered in the calculation of the freezing time. The conditions of the freezing medium, the temperature of freezing point, and the density are assumed to be stable during the freezing process. The thermal conductivity of the liquid phase, which is about a quarter of that of ice, is not considered. The effect of food composition is only expressed by the latent heat to be removed and not in the values for thermal conductivity of either frozen or unfrozen food. Few modifications are needed because of the assumptions increasing errors in estimation. Nevertheless, the most popular equation for freezing time was proposed by Planck (Keizer 1995; Tulek *et al.* 1999; Cemeroğlu & Soyer 2005).

According to the calculations, the power of the compressor for the freezer can be determined, so overload losses can be avoided. The temperature of the fish to be installed should be around 0°C, otherwise, the efficiency of the freezer decreases, the time of freezing extends, and spoilage before freezing is accelerated. The fish should be precooled.

The freezing time is affected by both characteristics of the food and the features of the freezing medium.

The chemical composition of fish will affect the freezing time. The fat content of fish increases with the decrease in water content. If there is less water, then less heat will require to be extracted to freeze the fish. Also, the coefficient of thermal conductivity of fish is an important factor affecting freezing time. The geometric shape, depending on the ratio of the surface area to the volume of the fish, is considered in many equations. The heat travelling distance to reach the surface from the inner part of the food affects the freezing time. The thicker the product, the longer is the freezing time.

The packaging and packaging material is one of the most important factors prolonging the freezing time. The heat transfer coefficient and thickness of the packaging material settle the effect. Air tapped between the packaging material and the food has a negative influence on freezing time. The type of the freezer, the temperature of the freezing process and the temperature of the fish are also important factors affecting freezing time (Gökoğlu 2002; Cemeroğlu & Soyer 2005).

7.1.7 Freezing velocity

Freezing time of a material cannot be a guide for freezing velocity. The ratio of the nearest distance between the central thermal point and surface of the material to the time needed for the temperature of that central point to decrease to 0°C is defined as freezing velocity. Freezing velocity is not stable and increases during freezing.

Rapid freezing is required to accomplish good-quality frozen products. Rapid freezing allows forming small ice-crystals reducing the tissue damage and hinders the leaking of fluids. Rapid freezing also provides the following benefits:

• Water is transformed into ice-crystals and where they are present they prevent water from entering the spaces between the cells.
• Formation of ice-crystals among the cells is provided, ensuring the protection of physical structure of the cells.
• The critical zone, −0.5/−5°C is passed rapidly.
• Microbiological degradation is inhibited by rapidly reaching unfavourable temperatures for microorganisms.

Recommended freezing and frozen storage temperatures for most perishable foods such as meat and fish are −35/−45°C and −20/−22°C, respectively. Rapid freezing in the freezing process of fish is achieved by reaching the temperature of the thermal centre from 0°C to −5°C less than 2 hours.

7.2 Biological aspects of freezing

7.2.1 Cryopreservation of cells and other biomaterials

Cryopreservation is defined as the storage of a living organism, or a portion thereof, at an ultralow temperature (typically colder than −130°C). At this low temperature, cellular viability can be stored in a genetically stable form upon thawing (Day & Stacey 2007; Sarder *et al.* 2012). Freezing will reduce thermal inactivation of the product and immobilize the solution components. Freezing preserves living cells and their components for long periods to avoid damage to the cells. Intracellular nucleation should be inhibited, because ice formation within a cell is lethal.

Among human tissues being cryopreserved are skin, corneal tissue, liver tissue, arteries and veins. Also, cryopreservation is an active field in preservation of semen, ova and embryos. It has a wide application in animal husbandry.

Fish stocks are threatened with pollution, overfishing and global warming. Cryopreservation of embryos from fish species is thought to be a chance for saving endangered species. Artificial fertilization in aquaculture has raised the requirement for storage of reproductive materials. Cryopreservation is a method suitable for the prolonged storage of fish sperm because they remain genetically stable and metabolically inert. In recent years, sperm of more than 200 species of fish have been successfully cryopreserved. Even though, in general, many successes have been achieved in fish sperm cryopreservation, the technique remains as a method that is difficult to be standardized and use in all types of fishes due to the fact that cryopreservation of sperms from different fish species required different conditions individually (Chew & Zulkafli 2012). Many studies have been performed on the cryopreservation of sperm of aquatic animals such as grayling, yellow catfish, cod, yellow drum, oyster, carp (Daly *et al.* 2008; Pan *et al.* 2008; Dai *et al.* 2012; Horvath *et al.* 2012; Tiersch *et al.* 2012; Yang *et al.* 2012).

In general, approximately 40–90% of spermatozoa from freshwater species are usually damaged after cryopreservation, whereas only 10–20% of spermatozoa are damaged in marine species. The difference between species is presumably a result of evolution of their cellular properties developed under the pressure of their niche environment. Freshwater species inhabit an environment of 0–50 mOsm, whereas marine fish live in a range of 600–1000 mOsm (Tiersch & Mazik 2003; Day & Stacey 2007).

Cryoprotectants are used to protect biological tissue from freezing damage. They lower the freezing point, promoting supercooling, protect cell membranes from freezing damage, and decrease the deleterious effect of high concentrations caused by freezing. The increase in solute concentration during freezing damages the cells as secondary effect of freezing. Sperm of marine species were successfully cryopreserved after the discovery of cryoprotectants.

The cryoprotectants should be biologically acceptable, be able to penetrate into the cells and have low toxicity. Many compounds have such properties, including glycerol, dimethyl sulfoxide, ethanediol and propanediol. The suitability of extenders and cryoprotectants differs among fish species. Some cryoprotectants, such as dimethyl sulphoxide (DMSO) (carp, European catfish) and methanol (catfish, salmonid), and comparisons of DMSO and methanol (sturgeon species) have been studied by many researchers (Lahnsteiner *et al.* 1997; Horvath *et al.* 2003, 2005; Christensen & Tiersch 2005; Linhart *et al.* 2005). The acute toxicity of methanol, 2-methoxyethanol (ME), DMSO, *N,N*-dimethylacetamide (DMA), *N,N*-dimethyl formamide (DMF), and glycerol with different concentrations were evaluated as cryoprotectants for the sperms of medaka fish (Yang *et al.* 2010). Similar study was performed with different concentration of methanol; MeOH, DMSO and DMA for the protection of the sperm of loach *Misgurnus anguillicaudatus* (Yasui *et al.* 2012). The success of cryopreservation strongly depends on the initial quality of the sperm.

7.2.2 Biological ice nucleation

Nucleation of ice is formed in two mechanisms. Only spontaneous aggregation of water molecules involve in homogeneous nucleation. The nuclei reach a critical size in the case of decreasing temperature and duration of chilling. The other mechanism, heterogeneous nucleation, is achieved with molecules other than water forming ice-crystals. Heterogeneous nucleation is more important in the freezing process. This type of nucleation occurs when water aggregates assemble on a nucleating agent with some insoluble material such as meat particles (Venugopal 2006). The formation of nucleation is required at temperatures below 0°C. However, the initial formation of ice is not kinetically favoured in pure water at temperatures above approximately −40°C. The heterogeneous ice nucleating agents of mineral and biological origin bind water molecules and prevent sudden and damaging ice formation. These ice-nucleating bacteria are able to initiate ice growth at temperatures as warm as −1°C (Lee & Costanzo 1998; Wowk & Fahy 2002).

Some microorganisms have the ability to catalyse ice nucleation. Some plant pathogenic bacteria species like *Pseudomonas syringae* and *Pseudomonas viridiflava*, and some saprophytic bacteria such as *Erwinia herbicola* and *Pseudomonas fluorescens* are ice-nucleation active strains. They promote the formation of ice-crystals owing to - ice nucleation proteins expressed by a single gene (Lindow 1997; Karel & Lund 2003).

The usage of ice-nucleation bacteria in freezing process of foods are reported to increase nucleation temperature, reduce freezing time, and improve the quality of frozen products. Also, it was reported that employing *Xanthomonas ampelina* TS206 strains in partial-freezing of shrimp at −3°C resulted in prolonged shelf-life (Zhang *et al.* 2010). *P. syringae* was sprayed to salmon muscle at a concentration of 10^7 colony-forming units (CFU)/ml and compared with the control group. It was found that the temperature difference between the freezing point temperature and nucleation temperature of the sample was reduced by 3.4°C (Li & Lee 1995).

7.2.3 Antifreeze proteins

Various fish species have the ability to survive at low temperatures owing to their proteins. These proteins depress the freezing point of body fluids and are found in many species of fish and invertebrates, mostly including insects, and in higher plants, as well as in fungi and bacteria (Griffith & Ewart 1995; Hassas-Roudsari & Goff 2012). Antifreeze proteins are characterized by their ability to prevent ice from growing upon cooling below the bulk melting point (Kristiansen & Zachariassen 2005).

Like most vertebrates, fish have an internal osmotic pressure of about 30–50% that of seawater. Without the aid of the antifreeze compounds their body fluids would rapidly freeze at the ambient temperature of −2°C. The biological antifreezes found in the fish prevent ice-crystals from propagating in their blood and body fluids allowing them to survive in cold waters (Cullins *et al.* 2011).

Based on the presence or absence of carbohydrates, they are classified into two main types: glycoproteins (AFGP) and non-glycoproteins (AFP). AFGP are primarily composed of repeating units of Al-Ala-Thr with glycosylation at the threonine residue. AFP are further subdivided into three distinct antifreeze protein subtypes: type I is alanine rich, found in flounders and sculpins, they range from 3.3 to 4.5 kDa in molecular mass; type II is cystine rich found in herring, and contains several S–S bonds, they range from 11 to 24 kDa in molecular mass; type III is found in eel pouts and wolf-fish, and are 6.5 kDa in molecular mass. The most recently discovered type IV is rich in glutamine in longhorn sculpin with a molecular mass of 12 kDa (Li & Sun 2002; Harding *et al.* 2003; Karel & Lund 2003).

These proteins function by suppressing the growth of ice nuclei by concentrating at the ice-crystal surfaces. It is generally accepted that antifreeze proteins function by binding to ice and interfering with water molecule propagation to the crystal surface. Thus, the freezing temperature decreases, recrystallization is inhibited and drip-loss is minimized during thawing.

Ice nucleation proteins raise the temperatures of ice nucleation and reduce the degree of supercooling, whereas antifreeze proteins can lower the freezing temperature and retard recrystallization on frozen storage (Li & Sun 2002). Antifreeze proteins and ice nucleation proteins, which are two functionally distinct and opposite classes of proteins, can be directly added to food and interact with ice, affecting the crystal size and structure (Hew & Yang 1992).

7.3 Freezing methods

The freezing process partially or completely hinders disruptive microbiological or enzymatic reactions, nevertheless, it cannot repair damage already caused. The fish should be frozen after handling for a better quality. The great advantage of the freezing process is the ability to achieve stability without damaging the initial quality. The material to be processed should be fresh and qualified, and handled in accordance with the technique of freezing in order to obtain long-lasting products.

Freezing is the process of heat transfer from the food to the cooling material, which can be gaseous, liquid or solid. Industrial equipment is designed to take account of this phenomenon. The method of generating the cooling effect and the method of applying by convection or conduction are the basic criteria.

Convection heat transfer is provided by transferring heat from one place to another with the movement of liquid (such as brine or cryogenic refrigerant) or gaseous (such as air). In conductive heat transfer, the material to be frozen needs to be in touch with the metal surface, which is directly cooled by a refrigerant or cooled secondarily by another medium.

Many parameters should be taken into account in choosing the right freezing methods such as shape, size and uniformity of the product, the rate of weight loss, rate of freezing, exclusion of oxygen during freezing, and the availability of the product to be packed.

The method should meet all financial, functional and feasible requirements (Keizer 1995).

7.3.1 Air blast freezing

Air is the medium that can be used for freezing for both packed and unpacked materials. The air is cooled by the evaporator of the refrigeration equipment and circulated by fans over the product. The velocity of the air should be 4–6 m/s to allow the acceptable heat transfer as well as keep the temperature of the medium at −30/−40°C. The performance of air freezers depends much on the velocity of air. Acceleration of air velocity is only effective on freezing of packed foods and increases the cost. This may cause freezer-burn and dehydration of unpacked fish unless they can be completely surrounded by flowing air.

There are different kinds of air blast freezing. Tunnel freezing is commonly used. The product is carried on a belt moving slowly through a tunnel or enclosure containing very cold air in motion. The direction of the flow can be parallel or in the opposite direction of the flow of the product. In the opposite flow tunnel, the product enters from one side of the tunnel, at the same time the cooled air flows from the other side. The coolest air meets with the product closest to frozen state and continues on its path through the tunnel. The freezing occurs gradually so as to not allow an increase in the temperature of the product. However, cool air heats up at the end of the tunnel. This warm air should return to the evaporator and needs to be cooled again. There will be a huge temperature difference between the air and the evaporator, promoting fast and continuous frost build up in the evaporator.

Another type of air blast freezing is called fluidized bed freezers. The products are placed on a mesh belt and exposed to forced air upward through the bed, sufficient enough to lift or suspend the particles. Each particle is surrounded by the cool air and freezes rapidly and individually. Fluidized bed freezers are a suitable method for individually quick frozen (IQF) products. The products needs to be small enough to take fluidity and have the same size and dimensions. Cool air and the product meet according to the principle of countercurrents, and the air is fed perpendicularly to the belt. Heat transfer between the product and the air is improved with provision of controlled turbulence. Weight loss of unpacked particles is decreased up to 0.6%. Small-sized seafood such as fish portions, fillets, shrimp, oyster and prawn are frozen with IQF technology and these products are the concern of several researchers (Jacobsen & Fossan 2001; Hatha *et al.* 2003; Boonsumrej *et al.* 2007; Songsaeng *et al.* 2010). This method of freezing allows the consumer to meet the exact quantity requirements to be removed from frozen storage.

Different shapes and size of fish can be frozen individually or in blocks, or packed or not hanging on coils, laid down on trolleys and plates, or placed in cabins and conveyor belts. Therefore, this method is widely used. However, the major problem of this method is moisture loss of unpacked products. Blowing high-velocity air among the products is the basic rule. The air has a drying effect, no matter how cold it is, depending on its humidity. In case of low air humidity, evaporation takes place regardless of temperature. The unpacked product will lose some moisture. Water joins the air as vapour and increases the humidity. Loss of water triggers two problems: the first is the physical change causing quality loss, depending on dehydration, and the second is the frost build-up on the evaporator spirals. The excessive dehydration and water loss through sublimation from the outer surface of the product leads to 'freezer-burn'. Freezer-burn decreases the quality related with appearance and causes nutrient loss. Also, lipid oxidation is triggered by the penetration of oxygen through porous structures provoked by freezer-burn. This event is irreversible. The initial step of the freezer-burn leads to a bright surface; then, however, dusky spots dominate due to oxidation. Dehydration also results in financial losses, depending on the amount of weight lost (Cemeroğlu & Soyer 2005).

Various measures can be taken in order to minimize dehydration and freezer-burn. Precooling with high humidity air at 4–5°C is an effective method of controlling dehydration. Packaging before freezing is a definite and important method. However, it is not correct for IQF and is costly for the other freezing methods because of delaying freezing time by reduction in heat transfer.

The application of a thin layer of ice on the already frozen fish prior to freezing loss is referred to as 'glazing'. Protection coating with ice excludes air from the surface of the product, reducing the rate of lipid oxidation and dehydration. Also, freezer-burn is retarded due to sublimation of the glaze instead of the tissue water. The layer of ice is formed by dipping the product into the glazing solution such as water, salt–sugar solutions or by spraying the glaze solution onto the product. It is important to adjust the degree of glazing. It is important to replace the glazing solution periodically in order to minimize the bacterial load and build-up of fish protein. Low degree of glazing (<6%) may lead to an unsatisfactory protection. Excessive glazing (>12%) might imply additional direct profits for sellers, which may lead to trade conflicts and misleading of consumers (Vanhaeckea *et al.* 2010). Goncalves and Junior (2009) demonstrated the effectiveness of the glazing process as a protecting agent for frozen shrimp and reported that reasonable range of water uptake could be between 15% and 20% to guarantee the final quality. The amount of glaze picked up by the product depends on the size and shape of the product, glazing time, the temperature of the seafood and glazing solution (Johnston *et al.* 1994; Jacobsen & Fossan 2001). It is important to refreeze the glazed product to avoid ice-crystal growth, tissue damage and drip-loss.

Chitosan solutions were compared against water as a glazing solvent for salmon fillets and it was found that chitosan coatings can be a good barrier to protect frozen fish from deterioration (Sathivel *et al.* 2007; Soares *et al.* 2013). Whey protein-based coatings were also studied for salmon (Rodriguez-Turienzo *et al.* 2011). Recently, antioxidant additives to glazing solutions have been studied. For this purpose extracts of tea, grape seeds and pomegranate peel were used and reported to obtain successful results in keeping the quality of fish (Lin & Lin 2005; Yerlikaya & Gökoğlu 2010).

7.3.2 Indirect contact freezing

The principle of this freezing method is separating the refrigerant of the cooling medium from the product to be frozen by conducting plates. These plates are often made of steel and their thermal conductivity is good. Each plate is equipped with evaporators. The heat transfer from the plates to the product is important. There should be an excellent contact on the heat exchange surface. The basic requirement in indirect contact freezing method is that the product should have a regular geometry, such as rectangular cartons, slabs such as fish sticks and fillets, or flat packages. Smooth-surfaced products placed side by side between two plates, which maintain pressure during the process, enforcing contact with the packages. A rapid freezing is provided on both sides.

Plate freezers are in batch mode or continuous mode. The batch plate freezers can be in horizontal or vertical arrangements containing many layers of plates. The feeding of the freezing system starts with the bottom layer in horizontal systems, whereas the loading starts from the upper layer in vertical systems. The frozen product is released on the other side of the freezer. The pressure of the plates against the product surfaces during freezing minimizes the bulging that often occurs in air-blast freezing systems, making the product squarer and easier to stack. Continuous plate freezers are commonly used for thin, flat-sided products with short freezing times, such as hamburger patties or fish fillets (Karel & Lund 2003; Cemeroğlu & Soyer 2005; North & Lovatt 2006).

7.3.3 Immersion freezing

In this freezing method, the food product is frozen by immersion in or by spraying with a freezant that remains liquid throughout the process. Direct contact with the food and a low-freezing-point liquid medium is provided. Cladding the product with the freezants results in rapid freezing, leading to a high-quality end product. The other advantages of this method are: amorphous products can be cooled, particulate products can be frozen individually, and oxidation-sensitive products can be frozen successfully because of cessation of the contact with the air.

The immersion solutions remain unfrozen at 0°C and below. The immersion solutions should not lead to changes in texture and taste of the food with which they are in direct contact. Also, the parameters such as cost, flavour compatibility, safety, and ability to reduce the solution freezing point, viscosity and thermal conductivity are considered in choosing immersion solutions (Lucas & Raoult-Wack 1998; Venugopal 2007). Therefore, the number of immersion solutions is limited: propylene glycol, glycerol and mixtures of salt and sugar. Combined solutions such as alcohol/water, water/glycerol/alcohol mixture are also utilized. The most widely used immersion solution is sodium chloride brine, which freezes at −21°C. The disadvantage of sodium chloride brine is that the fish may absorb some salt depending on the temperature of the brine, the immersion time, and the lipid content of the fish. The product must be wrapped in a sealed package to prevent direct contact of the brine with the food. It is difficult to maintain the purity and concentration of the immersion solution. Depletion of salt in the brine needs to be controlled with continual addition of salt.

Immersion freezing is commonly used for freezing tuna, shrimp and crab where whole fish or shrimp freeze in direct contact with refrigerated sodium chloride brine. Brine freezing has been used in large fishing vessels that fish a considerable distance from their home port and stay at sea for long periods (Aubourg & Gallardo 2005).

The immersion freezing technique can be assisted with power-ultrasound. The prolonged thermal effect upon the refrigerant is achieved by the continuous application of ultrasound. Acoustic treatment also shortens the time between the initiation of crystallization and the complete formation of ice, and reduces cell damage. This effect is ascribed to acoustic cavitation bubbles similar to the nuclei for crystal growth or disruption of nuclei already present. Moreover, ultrasound provides violent agitation, leading to an increase in heat and mass transfer. The acoustic energy can be applied in two ways; the first one is direct immersion of ultrasonic probes into the immersion solution, the second is indirect contact from a transducer coupling through parts of process vessels (Sun & Li 2003; Zheng & Sun 2006).

7.3.4 Cryogenic freezing

In this freezing method, the material to be frozen is exposed directly to liquid boiling or solid subliming at a very low temperature. Very rapid heat transfer is achieved (Karel & Lund 2003). Sensitive cell-structured foods such as strawberry, sliced tomato and mushroom can be frozen easily in accordance with the freshness criteria. This freezing method can be successfully utilized in seafood freezing.

A cryogen is the refrigerant that absorbs heat during phase transition. The coolants are liquid and solid CO_2, N_2, N_2O and Freon 12 (CCl_2F_2). Freon 12 was used in the 1960s and 1970s for freezing vegetables and shrimp. However, its use was abandoned completely due to its harmful effects on the environment. Liquefied gases with a very low boiling point are called cryogenic liquids. Low boiling point seems an advantage thermodynamically; however, this fact leads to mechanical damage in foods during freezing. Therefore, cryogenic freezants are requested to have boiling points around $-50°C/-60°C$ and a high latent heat of vaporization. The most employed food grade cryogenic freezants are liquid carbon dioxide and liquid nitrogen (Cemeroğlu & Soyer 2005).

Liquid carbon dioxide is acquired by compressing carbon dioxide gas under high pressure. The boiling point of liquid carbon dioxide is −145°C. The liquid is sprayed or dribbled onto the product while passing through a tunnel on a conveyor belt. The critical temperature, critical pressure and the temperature of the triple point of liquid carbon dioxide are 31°C, 7.35 MPa and −56°C, respectively (Gökoğlu 2002). Carbon dioxide is a solid at atmospheric pressure (called *dry-ice*) and sublimes at −79°C while absorbing 572 kJ/kg of heat.

Liquid nitrogen is obtained by compressing air and then separating the nitrogen gas from oxygen, by passing through a special valve. Liquid nitrogen remains liquid under atmospheric pressure and only a small amount is transformed into nitrogen gas at −196°C and absorbs 199.8 kJ/kg of heat during vaporization. Liquid nitrogen is utilized in cryogenic freezing process due to being non-toxic, inert, protects oxygenic reactions, has no need for extra cooling equipment, and ensures quick freezing at very low temperatures. Moreover, dehydration loss is minimized, product quality is ensured by quick and individual freezing, and the equipment is suitable for continuous flow operations.

The cryogenic freezing method is generally used for freezing shrimps. Cryogenic cooling is used in combination with other protection methods for improved quality and reduced cost. Cryogenic freezing is combined with air-blast freezing. It was reported that about 50% of the muscle cells surface cracking occurred in shrimps frozen at −120°C. Thus, freezing the tiger shrimps (*Penaeus monodan*) at −70°C was selected for the best condition of cryogenic freezing (Boonsumrej *et al.* 2007). Cryogenic freezing is also combined with sodium bicarbonate for freezing white shrimp (*Penaeus vannamei*) and found that this process improved yield, freezing time, freezing rate, cutting force and colour of shrimps (Lopkulkiaert *et al.* 2009). Guo *et al.* (2013) found that cryogenic freezing, ozonized water and antimicrobial coatings applied singly, achieved different degrees of microbial inactivation; however, their combinations provided additive or synergistic effects against naturally occurring bacteria and *Listeria innocua* on shrimp. Inactivation of *Salmonella* spp. by cryogenic freezing in combination with gamma radiation was studied by Sommers *et al.* (2011). Cryogenic freezing at −82°C of shrimp inactivated approximately 1.27 $\log_{10}$ of

Salmonella spp. Cryogenic freezing, in combination with a gamma radiation dose of 2.25 kGy, was sufficient to inactivate 5 $\log_{10}$ of *Salmonella* spp. on whole raw shrimp. Rodezno *et al.* (2013) compared the effect of cryogenic and air-blast freezing on catfish fillets and found that the cryogenically frozen catfish fillets had a lower freezing loss than fillets frozen by air blast freezing.

It is reported that the traditional freezing process is generally slow, resulting in large extracellular ice-crystal formation, which causes texture damage, accelerates enzyme activity and increases oxidation rates during storage and after thawing (Doughikollaee 2012). Novel methods are utilized in freezing processes. When water is frozen, its volume increases leading to a decrease in density. Under high pressure, the ice that is formed has a density greater than that of water. High-pressure ice does not expand in volume and high pressure maintains water in the liquid state. When the pressure is reduced rapidly, substantial supercooling is obtained (Karel & Lund 2003). The product quality is maintained by promoting uniform and rapid nucleation, formation of smaller ice-crystals and minimizing cell damage. High supercooling can be obtained by pressure treatments, allowing the formation of ice instantaneously and homogeneously throughout the product. High-pressure treatment can be combined with air-blast and liquid N_2 freezing methods (Knorr *et al.* 1998; Otero *et al.* 2000; Li & Sun 2002). Chevalier *et al.* (2000) compared the effect of pressure shift freezing (PSF, 200 MPa, −18°C), air-blast freezing and pressurized samples (200 MPa, 5°C) without freezing of whole Norway lobster. The effect of high pressure on myofibrillar proteins leads to increase in toughness with a decrease in salt soluble protein extractability. Lower pressures can be utilized to avoid the unfavourable effects on protein.

References

Abbas, K.A., Saleh, A.M., Mohamed, A. & Lasekan, O. (2009). The relationship between water activity and fish spoilage during cold storage: A review. *Journal of Food, Agriculture and Environment*, **7**(3–4): 86–90.

Alizadeh, E., Chapleau, N., de Lamballerie, M. & Le-Bail, A. (2007). Effect of different freezing processes on the microstructure of Atlantic salmon *(Salmo salar)* fillets. *Innovative Food Science and Emerging Technologies*, **8**: 493–499.

Aubourg, S.P. & Gallardo, J.M. (2005). Effect of brine freezing on the rancidity development during the frozen storage of small pelagic fish species. *European Food Research and Technology*, **220** (2): 107–112.

Belitz, H.D., Grosch, W. & Schieberle, P. (2009). Water. In: *Food Chemistry*, 4th revised and extended edition, pp. 1–7, Springer Press, Berlin.

Boonsumrej, S., Chaiwanichsiri, A., Tantratia, S., Suzuki, T. & Takai, R. (2007). Effects of freezing and thawing on the quality changes of tiger shrimp *(Penaeus monodon)* frozen by air-blast and cryogenic freezing. *Journal of Food Engineering*, **80**: 292–299.

Cemeroğlu, B. & Soyer, A. (2005). Gıdaların soğutulması ve dondurulması. *Gıda Mühendisliğinde Temel İşlemler*, pp. 1–358, Gıda Teknolojisi Derneği Yayınları. No:29, Ankara.

Chieh, C. (2006). Water chemistry and biochemistry. In: *Food Biochemistry and Food Processing* (Ed. Simpson, B.K.), pp. 103–133, Blackwell Publishing, Oxford, UK.

Chevalier, D., Sentissi, M., Havet, M. & Le Bail, A. (2000). Comparison of air-blast and pressure shift freezing on Norway lobster quality. *Journal of Food Science*, **65** (2): 329–333.

Chew, P.C. & Zulkafli, A.R. (2012). Sperm cryopreservation of some freshwater fish species in Malaysia. *Current Frontiers in Cryopreservation*, 269–292.

Christensen, J.M. & Tiersch, T.R. (2005). Cryopreservation of channel catfish sperm: effects of cryoprotectants exposure time, cooling rate, thawing conditions, and male-to-male variation. *Theriogenology*, **63**: 2103–2112.

Cullins, T.L., DeVries, A.L. & Torres, J.J. (2011). Antifreeze proteins in pelagic fishes from Marguerite Bay (Western Antarctica). *Deep-Sea Research II*, **58**: 1690–1694.

Cybulska, B. & Doe, P.E. (2002). Water and food quality. In: *Chemical and Functional Properties of Food Components* (Ed. Sikorski, Z.E.), pp. 25–49, CRC Press, Boca Raton.

Dai, T., Zhao, E., Lu, G., *et al.* (2012). Sperm cryopreservation of yellow drum *Nibea albiflora*: A special emphasis on post-thaw sperm quality. *Aquaculture*, **368–369** (24): 82–88.

Daly, J., Galloway, D., Bravington, W., Holland, M. & Ingram, B. (2008). Cryopreservation of sperm from Murray cod, *Maccullochella peelii peelii*. *Aquaculture*, **285** (1–4): 117–122.

Day, J.G. & Stacey, G.N. (2007). *Cryopreservation and Freeze-Drying Protocols*, pp. 347, Humana Press, New Jersey.

Doughikollaee, E.A. (2012). Freezing / Thawing and Cooking of Fish. In: *Scientific, Health and Social Aspects of the Food Industry* (Ed. Valdez, B.), pp. 57–67, InTech. Available from <http://www.intechopen.com/books/scientific-health-and-social-aspects-of-the-food-industry/epidemiology-of-foodborne-illness> (accessed 20 January 2015).

Erkan, N. & Varlık, C. (2004). Dondurarak muhafaza teknolojisi. *Su Ürünleri İşleme Teknolojisi*, pp. 96–127, Istanbul Universitesi Yayin no: 4465, Istanbul.

Gökoğlu, N. (2002). *Su Ürünleri İşleme Teknolojisi*, pp. 157, Su Vakfı Yayınları, Istanbul.

Goncalves, A.A. & Junior, C.S.G.G. (2009). The effect of glaze uptake on storage quality of frozen shrimp. *Journal of Food Engineering*, **90**: 285–290.

Griffith, M. & Ewart, K.V. (1995). Antifreeze proteins and their potential use in frozen foods. *Biotechnology Advances*, **13** (3): 373–402.

Guo, M., Jin, T.Z., Yang, R., *et al.* (2013). Inactivation of natural microflora and inoculated *Listeria innocua* on whole raw shrimp by ozonated water, antimicrobial coatings, and cryogenic freezing. *Food Control*, **34**: 24–30.

Harding, M.M., Erberg, P.I. & Haymet, A.D.J. (2003). Antifreeze glycoproteins from polarfish. *European Journal of Biochemistry*, **270**: 1381–139.

Hassas-Roudsari, M. & Goff, H.D. (2012). Ice structuring proteins from plants: Mechanism of action and food application. *Food Research International*, **46**: 425–436.

Hatha, A.A.M., Maqbool, T.K. & Kumar, S.S. (2003). Microbial quality of shrimp products of export trade produced from aquacultured shrimp. *International Journal of Food Microbiology*, **82**: 213–221.

Hew, C.L. & Yang, D.S.C. (1992). Protein interaction with ice. *European Journal of Biochemistry*, **203**: 33–42.

Horvath, A., Miskolczi, E. & Urbanyi, B. (2003). Cryopreservation of common carp sperm. *Aquatic Living Resources*, **16**: 457–460.

Horvath, A., Wayman, W.R., Urbanyi, B., Ware, K.M., Dean, J.C. & Tiersch, T.R. (2005). The relationship of the cryoprotectants methanol and dimethyl sulfoxide and hyperosmotic extenders on sperm cryopreservation of the two North American sturgeon species. *Aquaculture*, **247**: 243–251.

Horvath, A., Jesensek, D., Csorbai, B., *et al.* (2012). Application of sperm cryopreservation to hatchery practice and species conservation: A case of the Adriatic grayling *(Thymallus thymallus). Aquaculture*, **358–359** (15): 213–215.

Jacobsen, S. & Fossan, K.M. (2001). Temporal variations in the glaze uptake on individually quick frozen prawns as monitored by the CODEX standard and the enthalpy method. *Journal of Food Engineering*, **48**: 227–233.

Johnston, W.A., Nicholson, F.J., Roger, A. & Stroud, G.D. (1994). *Freezing and refrigerated storage in fisheries*. FAO Fisheries Technical Paper, 340, Food and Agriculture Organization, Rome.

Karel, M. & Lund, D.B. (2003). *Physical Principles of Food Preservation*, pp. 640, Taylor & Francis, New York.

Keizer, C. (1995). Freezing and chilling of fish. In: *Fish and Fishery Products* (Ed. Ruiter, A.), pp. 287–313, CAB International Press, Wallingford, UK.

Knorr, D., Schlueter, O. & Heinz, V. (1998). Impact of high hydrostatic pressure on phase transitions of foods. *Food Technology*, **52** (9): 42–45.

Kristiansen, E. & Zachariassen, K.E. (2005). The mechanism by which fish antifreeze proteins cause thermal hysteresis. *Cryobiology*, **51**: 262–280.

Lahnsteiner, F., Weismann, T. & Patzner, R.A. (1997). Methanol as cryoprotectants and the suitability of 1.2 ml and 5 ml straws for cryopreservation of semen from salmonid fishes. *Aquaculture Research*, **28**: 471–479.

Lee, R.E. & Costanzo, J.P. (1998). Biological ice nucleation and ice distribution in cold-hardy ectothermic animals. *Annual Review of Physiology,* **60**: 55–72.

Li, B. & Sun, D. (2002). Novel methods for rapid freezing and thawing of foods - a review. *Journal of Food Engineering,* **54**: 175–182.

Li, J. & Lee, T. (1995). Bacterial ice nucleation and its potential application in the food industry. *Trends in Food Science and Technology,* **6**: 259–265.

Lin, C. & Lin, C. (2005). Enhancement of the storage quality of frozen bonito fillets by glazing with tea extracts. *Food Control,* **16**: 169–175.

Lindow, S.E. (1997). Biological ice nucleation. In: *The Properties of Water in Foods* (Ed. Reid, D.S.), pp. 320–328, ISOPOW6, Springer, New York.

Linhart, O., Rodina, M., Flajshans, M., Gela, D. & Kocour, M. (2005). Cryopreservation of European catfish *Silurus glanis* sperm: sperm motility, viability, and hatching success of embryos. *Cryobiology,* **51**: 250–261.

Lopkulkiaert, W., Prapatsornwattana, K. & Rungsardthong, V. (2009). Effects of sodium bicarbonate containing traces of citric acid in combination with sodium chloride on yield and some properties of white shrimp *(Penaeus vannamei)* frozen by shelf freezing, air-blast and cryogenic freezing. *LWT - Food Science and Technology,* **42**: 768–776.

Lucas, T. & Raoult-Wack, A.L. (1998). Immersion chilling and freezing in aqueous refrigerating media: review and future trends. *International Journal of Refrigeration,* **21** (6): 419–429.

North, M.F. & Lovatt, S.J. (2006). Freezing methods and equipment. In: *Handbook of Frozen Food Processing and Packaging Facilities for the Cold Chain* (Ed. Da-Wen Sun), pp. 199–210, Taylor & Francis Group, CRC Press, New York.

Otero, L., Martino, M., Zaritzky, N., Solas, M. & Sanz, P. D. (2000). Preservation of microstructure in peach and mango during highpressure- shift freezing. *Journal of Food Science,* **65** (3): 466–470.

Pan, J., Ding, S., Ge, J., Yan, W., Hao, C., Chen, J. & Huang, Y. (2008). Development of cryopreservation for maintaining yellow catfish *Pelteobagrus fulvidraco* sperm. *Aquaculture,* **279** (1–4): 173–176.

Rodezno, L.A.E., Sundararajan, S., Solval, K.M., *et al.* (2013). Cryogenic and air blast freezing techniques and their effect on the quality of catfish fillets. *LWT - Food Science and Technology,* **54**: 377–382.

Rodriguez-Turienzo, L., Cobos, A., Moreno, V., Caride, A., Vieites, J.M. & Diaz, O., (2011). Whey protein-based coatings on frozen Atlantic salmon (*Salmo salar*): influence of the plasticiser and the moment of coating on quality preservation. *Food Chemistry,* **128**: 187–194.

Sarder, M.R.I., Sarker, M.F.M. & Saha, S.K. (2012). Cryopreservation of sperm of an indigenous endangered fish species *Nandus nandus* (Hamilton, 1822) for ex-situ conservation. *Cryobiology,* **65**: 202–209.

Sathivel, S., Liu, Q., Huang, J. & Prinyawiwatkul, W. (2007). The influence of chitosan glazing on the quality of skinless pink salmon *(Oncorhynchus gorbuscha)* fillets during frozen storage. *Journal of Food Engineering,* **83**: 366–373.

Soares, N.M., Mendes, T.S. & Vicente, A.A. (2013). Effects of chitosan-based solutions applied as edible coatings and water glazing on frozen salmon – a pilot-scale study. *Journal of Food Engineering*, **119**: 316–323.

Sommers, C.H., Rajkowski1, K.T., Sheen, S., Samer, C. & Bender, E. (2011). The effect of cryogenic freezing followed by gamma radiation on the survival of salmonella spp. on frozen shrimp. *Journal of Food Processing and Technology*, S8–001.

Songsaeng, S., Sophanodora, P., Kaewsrithong, J. & Ohshima, T. (2010). Quality changes in oyster *(Crassostrea belcheri)* during frozen storage as affected by freezing and antioxidant. *Food Chemistry*, **123**: 286–290.

Sun, D. & Li, B. (2003). Microstructural change of potato tissues frozen by ultrasound-assisted immersion freezing. *Journal of Food Engineering*, **57**: 337–345.

Tiersch, R. & Mazik, P. M. (2003). *Cryopreservation in Aquatic Species*, pp. 439, World Aquaculture Society, Baton Rouge, LA.

Tiersch, T.R., Yang, H. & Hu, E. (2012). Outlook for development of high-throughput cryopreservation for small-bodied biomedical model fishes. *Comparative Biochemistry and Physiology Part C: Toxicology & Pharmacology*, **155** (1): 49–54.

Tulek, Y., Gokalp, H.Y. & Ozkal, S.G. (1999). Gıdaların donma ve çözülme zamanlarının belirlenmesinde kullanılan tahmin metotları I. basit eşitlikler. *Journal of Engineering Sciences*, **5** (1): 943–950.

Vanhaeckea, L., Verbekeb, W. & De Brabander, H.F. (2010). Glazing of frozen fish: Analytical and economic challenges. *Analytica Chimica Acta*, **672**: 40–44.

Venugopal, V. (2006). Quick freezing and individually quick frozen products. *Seafood Processing*, pp. 446, Taylor and Francis, CRC Press, New York.

Wowk, B. & Fahy, G.M. (2002). Inhibition of bacterial ice nucleation by polyglycerol polymers. *Cryobiology*, **44**: 14–23.

Yang, H., Norris, M., Winn, R. & Tiersch, T.R. (2010). Evaluation of cryoprotectant and cooling rate for sperm cryopreservation in the euryhaline fish medaka *Oryzias latipes*. *Cryobiology*, **61** (2): 211–219.

Yang, H., Hu, E., Cuevas-Uribe, R., Supan, J., Guo, X. & Tiersch, T.R. (2012). High-throughput sperm cryopreservation of eastern oyster *Crassostrea virginica*. *Aquaculture*, **344–349** (21): 223–230.

Yasui, G.S., Fujimoto, T., Arias-Rodriguez, L., Takagi, Y. & Arai, K. (2012). The effect of ions and cryoprotectants upon sperm motility and fertilization success in the loach *Misgurnus anguillicaudatus*. *Aquaculture*, **344–349**: 147–152.

Yerlikaya, P. & Gökoğlu, N. (2010). Effect of previous plant extract treatment on sensory and physical properties of frozen bonito *(Sarda sarda)* filllets. *Turkish Journal of Fisheries and Aquatic Sciences*, **10**: 341–349.

Zhang, S., Wang, H. & Chen, G. (2010). Addition of ice-nucleation active bacteria: *Pseudomonas syringae* pv. *panici* on freezing of solid model food. *LWT- Food Science and Technology*, **43**: 1414–1418.

Zheng, L. & Sun, D. (2006). Innovative applications of power ultrasound during food freezing processes-a review. *Trends in Food Science and Technology*, **17**: 16–23.

CHAPTER 8

Freezing and frozen storage of fish

8.1 Effects of freezing and frozen storage on fish quality

Biochemical and physical changes are retarded by the freezing process. Although being one of the most effective preservation methods, freezing cannot eliminate all undesirable changes in foods. Deterioration of fish during frozen storage depends on many factors, including fish species, freezing rate, temperature, time of frozen storage, freezing method utilized and enzymatic degradation. Fish muscle undergoes chemical and structural changes during freezing and frozen storage. The most important changes are oxidation of lipids resulting in rancid odour and flavour, toughening due to protein denaturation and aggregation, discoloration largely due to oxidation reactions, and freezer-burn during freezing and frozen storage (Kolbe & Kramer 2007).

The conversion of water to ice during the freezing process increases the concentration of dissolved materials. The increased concentration causes a change in acid–base equilibrium. Shifts in pH of up to 1 pH unit, usually towards the acid side, may be observed. The salts, and other compounds such as phosphate, precipitate and result in drastic pH changes up to 2 pH units. These changes affect the physicochemical

Seafood Chilling, Refrigeration and Freezing: Science and Technology, First Edition.
Nalan Gökoğlu and Pınar Yerlikaya.
© 2015 John Wiley & Sons, Ltd. Published 2015 by John Wiley & Sons, Ltd.

food systems irreversibly. Loss of water-holding capacity, removal of enzymes from cell particles, actomyosin changes in muscle, and the alterations related with these changes are the effects of freezing process (Hui *et al.* 2004). Biochemical indicators of deterioration during frozen storage include: (1) protein denaturation (extractability, hydrophobicity, viscosity and electrophoretic patterns), (2) decrease or increase of enzyme activity, and (3) change in metabolite concentrations (amines, aldehydes and nucleotide degradation) (Sun 2006). Oxidation of lipids and proteins are also important factors affecting the quality of a frozen product.

8.1.1 Chemical and nutritional changes

Fish muscle consists of dark and light muscles. Dark muscle is used for sustained activity. The swimmers like tuna and mackerel have greater amounts of dark muscle, whereas light muscle is present in all fish species. The sarcoplasmic proteins, which are rich in pelagic fish, include enzymes and oxygen carriers. Myosin, actin, tropomyosin and troponin are myofibrillar proteins that are responsible for muscle contraction.

Frozen storage results in freezing denaturation of fish proteins. This alteration takes place especially in myofibrillar proteins. The myofibrillar proteins form protein–protein bonds causing high-molecular-weight polymers to form and become unextractable in salt solutions. The increase in salt concentration of the unfrozen phase contributes to the denaturation of protein in association with physicochemical changes during frozen storage. The insolubility of protein is a serious problem than lipid oxidation in lean fish. The progress of protein insolubility depends on fish species, nutritional status, rigor stage, pretreatments, freezing velocity, and lipid oxidation and storage temperature. The degree of protein denaturation is lower in freezing in the pre-rigor stage than in the post-rigor stage of fish. Denatured proteins also lose their enzymatic activity. Secondary reactions take place between protein and different reactants. The reason of secondary interactions is lower extractability and reduced functionality of myofibrillar proteins during frozen storage (Benjakul *et al.* 2003). It was reported that marine animal muscles are more susceptible to protein denaturation by freezing than mammalian muscle (Erkan & Varlik 2004; Diaz-Tenorio *et al.* 2007). The changes in the functional properties of frozen fish muscle have been attributed to conformational transitions of muscle proteins, that lead to protein aggregation, involving hydrophobic interactions, hydrogen bridges and the formation of covalent, non-disulphide bonds, as well as to changes in the structure of the water, and/or alterations in the protein–water interactions, together with a transfer of water to larger spatial domains (Sánchez-Alonso *et al.* 2012).

The reason for protein denaturation during freezing or frozen storage is considered to be the salt concentration and denaturing due to freezing out of water; water molecules surrounding the protein molecules change the conformation of the protein, and free fatty acids formed affect myofibrillar protein stability (Kolbe & Kramer 2007). Many functional properties of proteins have been related to solubility or extractability of the myofibrillar fraction. Freezing, particularly at a slow rate, initially results in a noticeable decrease in myosin–actin affinity due to denaturation of the myosin heads. This is followed by a continued decrease in ATPase activity and myosin tail denaturation (Sun 2006). Frozen storage directly affected the conformational changes in protein molecules, leading to the loss in functionality such as water holding capacity, viscosity, gel forming ability and lipid emulsifying capacity. Undesirable odour, consistency, hardness and stickiness are observed in frozen products because of protein denaturation.

The enzyme that causes trimethylamine oxide (TMAO) breakdown to dimethylamine (DMA) and formaldehyde (FA) is active down to as low as −8°C. Formaldehyde level in fish muscle has been used as an index of frozen storage deterioration. In formaldehyde-producing species such as gadoids the texture and functionality of the muscle deteriorate faster during frozen storage than they do in non-formaldehyde-forming species. This has been associated with DMA and FA formation during frozen storage of these species, which is generally accompanied by a decrease in extractability of myofibrillar proteins (Tejada *et al.* 2003). FA has been known to react with different functional groups of protein side chains and form intra- and intermolecular methylene bridges. Myofibrillar proteins that react with FA become denatured and the protein aggregate occurs (Benjakul *et al.* 2005). FA accumulation is higher in dark muscles of the fish. The free bound water in these kinds of products is held together with loose ties. This water is separated during the first chewing. The dry and cottony texture of the fish can be sensed at later stages. It was reported that FA participates in the formation of interaction compounds with fluorescent properties (Aubourg 1998).

The proportion of disulphide and non-disulphide covalent bonds with protein also increases with frozen storage time. Denaturation and aggregation of muscle proteins are associated with the formation of disulfide, as well as the formaldehyde formation (Benjakul *et al.* 2003). Continuous decrease in sulphydryl group with a concomitant increase in disulfide bond formation in lizardfish, croaker, threadfin bream and bigeye snapper was reported during extended frozen storage (Benjakul *et al.* 2000).

Proteins are also known to be susceptible to oxidative reactions. When proteins are exposed to oxygen, amino acids are destructed and protein-lipid bonds form. Protein–lipid aggregates are structured between carbonyl groups of oxidized lipids and protein molecules. These reactions result in the formation of protein-centred radicals (Saeed & Howell 2002). Modification of amino acid residue side chains, formation of protein carbonyls, cleavage of the peptide bonds, formation of covalent intermolecular cross-linked derivatives, loss of sulfydryl groups are the effects of oxidation on protein molecules. There are several impacts of protein oxidation on muscle quality such as changes in solubility and protein functionality, loss of essential amino acids and decreased digestibility (Lund *et al.* 2011).

During the frozen storage of fish, lipid hydrolysis and oxidation occurs influencing protein denaturation, texture changes, functionality loss and fluorescence development. Lipid oxidation compounds react with proteins, peptides, free amino acids and phospholipids (Aubourg 1999; Aubourg & Medina 1999). Hydrophobic effect of free fatty acids on proteins, interaction of oxidized lipids with cystine-SH, the epsilon-NH_3 group of lysine and N-terminus groups of aspartic acid, tyrosine, methionine and arginine of fish proteins are reported to influence the solubility of myofibrillar, and sarcoplasmic proteins (Siddaiah *et al.* 2001).

Highly unsaturated fatty acids are prone to lipid oxidation. These poly-unsaturated fatty acids, which are rich in fish lipids, are oxidized to hydroperoxides in the presence of oxygen and pro-oxidant molecules such as heme pigments and trace elements. The rate of oxidation increase with increased amount of unsaturation and double-bonds of fatty acids, temperature and light. In further oxidation, hydroperoxides are decomposed into multitude compounds. Non-volatile and volatile compounds, of different molecular weight and polarity, and bearing different oxygenated functions, such as hydroperoxy, hydroxy, aldehyde, epoxy and ketone functions, are formed (Dobargenes & Marquez-Ruiz 2003; Gimenez *et al.* 2011). Aldehydes can form cross-links with different compounds such as proteins leading to hardening muscle tissue. Malondialdehyde and gluteraldehyde interact with other compounds such as amines, nucleosides and nucleic acids, proteins, amino acids and phospholipids, or other aldehydes that are end-products of lipid oxidation, to form polymers resulting in structural damage and changes in functionality (Pereira de Abreu *et al.* 2010, 2011). The storage temperature of frozen fish should be as low as possible to inhibit oxidation of the highly unsaturated lipids directly related to the production of off-flavour. The oxidation of lipids leads to undesirable changes in taste, odour and texture, functionality of the muscle and a reduction in nutritional quality. Fatty fish has a shelf life of 4–6 months at −18°C, whereas lean fish can be preserved 7 to 12 months (Hui *et al.* 2004).

Endogenous fish enzymes may be active during frozen storage, even at −20 °C (Hwang & Regenstein 1989). Frozen storage temperatures lead to lysosomal enzyme leakage because the lipid peroxidation enzyme system is active at temperatures below the freezing point of fish tissue. It was suggested that phospholipase may be activated by freezing and it would be possible that enzymatic lipid peroxidation activates phospholipase to initiate phospholipid hydrolysis in frozen fish muscle. Lipases and phospholipases are the main factors in the hydrolysis and formation of free fatty acids. Both fatty and lean fish species show significant lipid hydrolysis during frozen storage. The increase in fatty acids cause rancid flavour and deleterious effects on ATPase activity. Free fatty acids are known to cause texture deterioration by interacting with proteins and strongly interrelated with lipid oxidation Free fatty acids oxidize more rapidly to form higher-molecular-weight lipids (triglycerides and phospholipids), thus providing greater accessibility to oxygen molecules and other pro-oxidant molecules (Aubourg 1999; Sun 2006; Aubourg *et al.* 2007; Barthet *et al.* 2008).

Freezing can disrupt muscle cells, resulting in the release of mitochondrial and lysosomal enzymes into sarcoplasm. Frozen products lose water-soluble proteins, vitamins and minerals during thawing, with leaking of the expressible fluid. These losses increase with the increasing proportion of denatured proteins. Lipid oxidation affects the nutritional value of the frozen food. The formation of lipid oxidation products, such as hydroperoxides, triggers oxidative interactions with sulphur-containing proteins, resulting in nutritional losses (Cemeroğlu 2005). The oxidation of proteins leads to the loss of essential amino acids and decreases the digestibility of the product.

8.1.2 Microbiological changes

The freezing process is a physical preservation method keeping the properties of fresh fish more than any other preservative process. The free water present in the structure turns into ice-crystals, leading to decrease in water activity and temperature. Ice-crystals are not convenient for the microorganisms and simulate the effect of drying. The increase in osmotic pressure of the concentrated unfrozen solution prevents microbiological growth. Ensuring the persistence of the state of being frozen is the most important factor in controlling biochemical and chemical reactions. The low temperature has a contributory effect on the prevention of microbial growth. The functions of microorganisms are terminated exposing to certain temperatures. Most of the pathogenic bacteria are inactivated in the temperatures below −10°C. Therefore, this temperature is convenient for frozen storage in order to prevent the spoilage of foods. Some of the microorganisms are able to grow even below −18°C. Storage time of the food extends with the decreasing temperature. However, temperatures below −18°C have marginal effects for most of the products because of the energy costs.

Freezing inhibits the activity of food spoilage and food-poisoning organisms. The targets of the freezing process are: (i) inhibition of the growth and molecular activity of microorganisms by reducing temperature; and (ii) removal of water by turning it into ice. The water available for the maintenance of microorganisms decreases during freezing of fish. The water activity of the food decreases with the decreasing temperature. The microorganisms are exposed to low water activity, low temperature, increasing solute concentrations and substantial pH changes in the freezing process (Mazur 1966). Microbiological activity in frozen fish is negligibly small or completely stopped. Low temperatures retard bacterial activity, allow a steady die-off, but do not intercept the deterioration of food entirely. The intra- and extracellular ice-crystals induce irreversible damage to bacterial cell membranes (Duan *et al.* 2010). During frozen storage, microorganisms may be damaged by physical or chemical reactions between cell components and components of the surrounding medium, or by desiccation of food surfaces (Gill 2006).

A general reduction of the bacterial populations in seafood is determined for pathogens as well as for psychrophilic spoilage organisms. Generally, Gram-negative pathogens such as *Salmonella* and other Enterobacteriaceae are sensitive to freezing damage. Spores are unaffected by freezing, and vegetative cells of Gram-positive bacteria, including *Staphylococcus* and *Listeria*, usually survive (Pan & Chow 2004).

It has been reported that the lethal effect of temperature on microorganisms was highest between −4°C and −10°C rather than lower temperatures. The most sensitive bacteria are Gram-negative bacterial vegetative cells, and the most resistant are spores and Gram-positive bacteria. However, when the food is thawed the surviving bacteria from the freezing and frozen storage still have a chance to grow and multiply (Kolbe & Kramer 2007). *Aeromonas* spp. are able to survive and multiply at low temperatures and produce virulent factors, even at these low temperatures. The incidence of *Aeromonas* spp. was detected in frozen freshwater fish intended for human consumption (Castro-Escarpulli *et al.* 2003). In another study, channel catfish *Ictalurus punctatus* were injected with *Aeromonas hydrophila, Pseudomonas fluorescens, Edwardsiella tarda, E. ictraluri* and then frozen at −20°C. *A. hydrophila, P. fluorescens, E. tarda* and *E. ictaluri* were isolated 20, 60, 50 and 30 days, respectively, after frozen storage (Brady & Vinitnantharat 1990). The microbiological load of the food mostly declines or cannot be detected after extended frozen storage.

Fish parasites are part of the ecosystem and can be found in the sea and fresh water naturally. To avoid these parasites, fish needs to be cooked, reaching an internal temperature of 70°C. Freezing the fish at appropriate temperatures and durations will kill the parasites and eliminate the possibility of infection. The parasitic nematode and cestode worms of public health concern are sensitive to freezing. Larvae of the various helminths that are present in fish can be inactivated by freezing methods. Freezing fish is a kind of control measurement when processing a raw or lightly preserved fish. Freezing as a means of killing parasites is time/temperature dependent. Generally, for parasitic worms, 15 hours in a blast freezer at −35°C or 7 days at −20°C will be effective (Adams *et al.* 1997). The lipid content of the fish and the type of the parasite present also affect the effectiveness of freezing to kill parasites.

The microorganisms that survive during frozen storage resume growth after thawing and may lead to spoilage. Psychrophilic or psychrotrophic microorganisms can grow at cold thawing temperatures. The survival rate of microorganisms depends on the natural microflora, the type of microorganism, the stage in the life cycle of the microorganism, the pre-freezing fish quality, the freezing rate, the storage temperature and fluctuations, and the method of thawing.

8.1.3 Physical changes

The freezing rate is critical to the nucleation and growth of ice-crystals, their size and location. These aspects are considered to be effective on the textural quality of frozen food. The traditional freezing process is generally slow, resulting in large, damaging extracellular ice-crystal formations. The exterior cells cool rapidly and reach the initial freezing point. Water separates from the solute and the salt concentration in the extracellular fluid increases. This phenomenon leads the formation of extracellular ice-crystals, which stretch the tissues resulting in textural damage, accelerate enzyme activity and increase oxidation rates during storage and after thawing. Conversely, rapid freezing can reduce ice-crystal size due to minimizing the migration of water into the extracellular space, but it may cause lethal intracellular ice-crystallization in living cells and mechanical cracking (Chevalier *et al.* 2001; Alizadeh *et al.* 2007). Structural damage of muscle during slow freezing is greater than fast freezing.

The surface hydrophobicity, amino acid modification, conformational stability, solubility and aggregation of the muscle proteins affects the textural characteristics such as water-holding capacity, viscosity, gelatin, emulsification, foaming and whipping (Xiong 1997). The denaturation and cross-linking of proteins leads the fish muscle to be tough, dry and fibrous. In the advanced stages of denaturation, thawed fish muscle seems matt and opaque and becomes spongy structure (Gökoğlu 2002). These properties are related to low water-holding capacity, a decrease in extractability and solubility of myofibrillar proteins, and an increase in the shear strength.

Freezing is a kind of dehydration process in which frozen water is removed from the original location to form ice-crystals in food. During thawing, the tissue does not have the ability to reabsorb the melted ice-crystals. This leads to undesirable release of exudates (drip-loss) and toughness of texture. Myosin and actin are largely responsible for the functional properties of muscle. Myosin from fish is generally less stable than its mammalian or avian counterparts. This characteristic is manifested in the stability of the flesh on frozen storage. During frozen storage, myosin in particular, undergoes aggregation reactions, which lead to toughening of the muscle and a loss in water-holding capacity (Mackie 1993; Sigurgisladottir *et al.* 2000). The decrease in water-holding capacity in fish muscle due to fibre shrinkage, cell damage, lower protein solubility, protein denaturation, and aggregation taking place during freezing and thawing, all lead to drip-loss. Soluble nutrients and flavour compounds are lost with expressible fluid from fillets. Also, the appearance of the food seems unpleasant after drip-loss. The muscle fibres of the fish reabsorb most of the melt water in the case of being stored at low temperatures, avoiding fluctuations during frozen storage. Frozen Atlantic salmon, compared to unfrozen pre-rigor fillets, had a higher drip-loss during cold storage, which shows that freezing reduces quality of pre-rigor fillets (Einen *et al.* 2002). In contrast, freezing of pre-rigor compared to post-rigor fillets of white fish species is known to increase drip-loss (Sørensen *et al.* 1997). This shows that extrapolating results with regard to yield and drip-loss from lean fish species such as cod to fat fish species such as salmon can be misleading.

Moisture loss during frozen storage is a serious problem, depending on the temperature of the evaporator and temperature fluctuations. The damage from dehydration and oxidation of food during cold storage causes freezer-burn. In this case, water molecules in frozen foods migrate to the surface and the layer of ice sublimes away. Ice-crystal evaporation from the surface results in freezer-burn. Thin parts of the fish such as fins and tail ends are sensitive to this. Freezer-burn affects the colour, texture and flavour of frozen foods such as dry spots, opaque dehydrated surface, and porous and spongy structure. Freezer-burn does not mean that the food is unsafe, only less desirable. The product needs to be protected by appropriate package or thin layer of ice (glazing). Fluctuations in temperature within a freezer also contribute to the onset of freezer-burn. The temperature difference between the air and the refrigerant should be as small as possible in order to limit dehydration and the need for defrosting the air coolers.

The term 'gaping' occurs when the connective tissue in fish muscle (myocommata) fails to hold the muscle segments or blocks together. The surface of a fillet looks split or cracked. The gaping is often associated with low final pH of the muscle and the changes in connective tissues during freezing. To avoid shrinkage the fish should be kept chilled and frozen before the pre-rigor stage. In case of freezing in the pre-rigor stage, the rigor progress will probably have completed during frozen storage and defrosting (Venugopal 2006). Pre-rigor filleting allows muscle fibres to contract freely from the vertebrae, which means lower tension between muscle fibres and myocommata, leading to less gaping (Einen *et al.* 2002).

The fish are classified according to their fat content: lean fish such as cod, haddock, saithe and plaice; semifatty fish such as halibut, rainbow trout and wolf-fish; and fatty fish such as salmon, mackerel, herring and eel. The changes in texture and functionality observed in the muscle of fatty and semifatty species are smaller than in lean species (Tejada *et al.* 2003).

8.1.4 Sensory changes

Sensorial quality is an important characteristic for consumer judgement as an indicator of food quality. The fresh fish is succulent and has desirable odour and flavour. The correct level of temperature, humidity, and air circulation should be selected to prevent excessive shrinkage, toughening, and discolouration. If the fish is stored in improper conditions, it becomes opaque, dull, and spongy when thawed, and the flesh may lose integrity.

One of the biochemical processes affecting texture, flavour and colour is protein denaturation. Lipid–protein cross links decrease protein solubility and contribute to the formation of coloured complexes. The pigments of the muscle exposed to colour deteoriations. Colour influences the acceptability of the consumer as an indicator of the product quality. The colour of fish changes after freezing and thawing processes. The changes in the rigor contraction and muscle texture during freezing affects the light reflectance properties and the visual impression of fillet colour (Einen *et al.* 2002). The colour of fish lipids are caused by lipid-soluble pigments. These pigments are oxidized during freezing and frozen storage. The red colour of shrimps fades during frozen storage and the shrimps gradually appear more yellow due to photo-oxidation of astaxanthin (Bak *et al.* 1999). Freezing and thawing promote oxidation of carotenoids in salmon fillets. It was reported that frozen storage caused an increase in luminosity, redness and yellowness of raw and smoked fillets of salmon (Regost *et al.*, 2004). A high content of conjugated double bonds exposed to oxygen causes colour loss in carotenoids. In tuna, myoglobin turns into metmyoglobin because of oxidation. Similarly, salmon loose its pink colour due to oxidation of carotenoids. In crab and lobster, 'blueing' discoloration may occur during freezing or frozen storage due to being in contact with oxygen (Cemeroğlu 2005). Swordfish develops a green colour beneath its skin due to an oxidation product; sulph-haemoglobin during frozen storage (Jeremiah 1996).

Freezer-burn is caused by air reaching to the food surface in means of dehydration and oxidation. Evaporation of water continues during freezing leading to form dry spots and colour changes. The fish needs to be glazed and temperature fluctuations need to be minimized in order to lessen freezer-burn damage. Another solution to avoid freezer-burn and dehydration is packaging. The material of the package should be thick enough to protect the product, and also thin enough to ensure the penetration of low temperatures.

The alteration of fish odour is composed of two stages: first, losing the characteristic fish odour and second, formation of undesirable odour. Generally, marine-derived foods have the odour of fresh seaweed. The fish loses its intense fresh odour and flavour after freezing. The oxidation of lipids, enzyme-catalysed changes and the degradation products of TMAO result in rancid odours and flavours. Acid and carbonyl compounds are formed during oxidation of lipids and pigments that have unpleasant odours and flavours.

8.2 Shelf life of frozen fish

The overall quality of a frozen fish depends on many factors, such as the condition of the fish before freezing, pretreatments, the freezing method, and temperature and humidity during storage and transportation. Marine-derived products are very perishable and need to be transferred to the consumer as fast as possible. The maximum quality is maintained by fresh-chilling, freezing and ensuring the cold chain is maintained. The quality of the final product mainly depends on the initial quality, pretreatments, processing, packaging, storage time and temperature.

The shelf life of a frozen fish is characterized by the initial quality, the freezing method, the conditions during frozen storage and also the fish species, catching season, spawning period, glazing, packaging, etc. The quality of a frozen fish and its stability during frozen storage depends on the storage temperature and the fluctuations of the temperature. Temperatures below −40°C allow the fish quality to be maintained for 6 to 9 months. These temperatures inhibits the protein denaturation and lipid oxidation, however cannot be utilized due to economical concern.

The important thing in the frozen storage of fatty fish is the oxidative changes in lipids and pigments affecting the odour and colour as well as proteins. Meanwhile, the main change in lean fish is aggregations of proteins related to the texture of the muscle (Badii & Howell 2002). Fatty fish such as mackerel, somon and ringa have a shelf-life of 2–3 months at −18°C, whereas the quality of lean fish such as cod and sole can be stable for long periods at the same temperature. The cold storage life of seafood is given in Table 8.1. It is recommended that fish prone

Table 8.1 The cold storage life of seafood*

Seafood	Refrigerated storage	Frozen storage
Lean fish (cod, flounder, haddock etc.)	1–2 days	6–10 months
Fatty fish (bluefish, salmon, mackerel etc.)	1–2 days	2–3 months
Smoked fish	2 weeks	4–5 weeks
Clams, crab, lobster in shell	2 days	3 months
Oysters and scallops	4–5 days	3–4 months
Shrimps	4–5 days	3 months
Shucked clams	4–5 days	3 months
Tuna salad, store prepared or homemade	3–5 days	

*Boyer & McKinney (2013).

to lipid oxidation needs to be stored at –29°C, while the others can be stored between –18°C and –23°C. The fish species in which textural changes occurs due to TMAO degradation need to be stored at temperatures below –30°C (Cemeroğlu 2005).

8.3 Freezing of fish on board

The quality changes of fish starts immediately after catching. The preservation of fish starts on the vessel as soon as it has been caught. Quick freezing and low-temperature storage is the most appropriate process to keep the initial quality of the fish in prolonged catching in deep-fishing. Freezing at sea has a crucial role in fish processing. The reason of freezing at sea is reaching a final product as good as that freshly caught and certainly better than an iced product held several days. A well-treated frozen product ready for shipment enables the fisherman to hold the product until some later time when the price is up (Kolbe & Kramer 2007).

The first thing after catching should be separating the non-food materials and the materials that may adversely affect the odour and flavour of the basic product. The fish needs to be moisturized and chilled with cold sea water or ice. The fish is generally eviscerated, bled and washed at once. The removal of the gut releases blood from the fish. This process may differ from fisherman to fisherman. The fish can be frozen without any preliminary process such as eviscerating or filleting. The whole fish, when thawed after landing, are available for any form of traditional processing. The fish going into rigor is a serious problem in large catches. The fish should never be frozen while in rigor stage as this will result in broken, gaping fillets on defrosting (Garthwaite 1997).

The early catch should be frozen primarily. Either whole fish or fillets can be frozen in plate freezers or blast on ships. Therefore, the fish should be classified according to their size. The shape of the fish should be neat in case of being in rigor stage. Otherwise, the flesh of the fish is fractured in the freezing process. The fish loins are placed on trays in the blast freezer or hung from racks (Figure 8.1). Generally, horizontal plate freezers are used for fillets, while vertical plate freezers are used for round-gutted fish. The fish needs to be

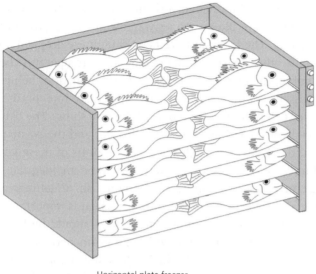

Horizontal plate freezer

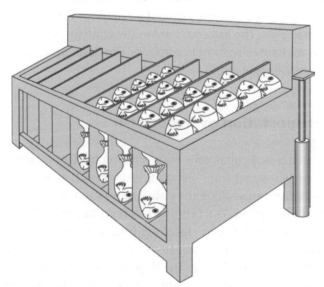

Vertical plate freezer

Figure 8.1 Freezing fish on board.

placed head to tail while loading into the plate freezer ensuring the head is towards the outside of the block to reduce any damage (Garthwaite 1997).

Brine freezing of individual fish like tuna can be carried out on board. The tanks are filled with seawater and cooled around 0°C just before catching. The fish that have been caught are dropped into these tanks and the salt is added in order to reduce the temperature below the freezing point. The fish are removed from the tank after being frozen.

The vessels having large capacity and high rate of fishing are designed for fish processing and named 'factory ships'. The fish may be frozen whole or filleted, packaged and frozen, and the waste products converted to fish meal and oil. The freezing units of these vessels are adjusted to catching capacity, so there should be an adequate amount of storage in order to separate the caught fish. Cleaning, washing, precooling and cold storage rooms are needed. It is important to avoid forming fish stacks and crushing. The freezer should be loaded with only the amount that will freeze within 24 hours because overloading slows down the freezing rate, leading to loss of fish quality. The freezer should be able to operate with part loads. The deck needs to be easily cleaned. Appropriate materials should be used for fish stock areas and the tank materials. The material used in the construction of the freezer should be resistant to seawater corrosion. The drainage systems should be designed (Gökoğlu 2002).

8.4 Transportation of frozen fish

The fish should be kept as cool as possible immediately after catching in order to withstand long shipping distances. It is important that chilled conditions exist along the entire transportation and production line until consumption. The fish that has been frozen on board should be transferred to the frozen stores on land as soon as possible. It is better to locate cold stores of fish stocks close to the landing place. The transferring of frozen fish from the ships to the shore should be done immediately. In case of delay, the frozen fish may thaw partially, resulting in loss of quality. The frozen fish should not be exposed to sunlight.

Various types of fish such as whole, gutted, dressed (having head and guts, but tails and fins may be removed), individually frozen fish or fish blocks, fillets individually frozen, or frozen blocks of fillets are transported. Frozen fish needs to be transported in refrigerated or frigorific insulated vehicles capable of maintaining the temperature of approximately −18°C over long distances. Reefer ships or vessels are specialized cargo vessels that are able to carry any frozen and cooled product that needs to be maintained at temperatures other than ambient temperature. These vehicles are designed as cold stores to maintain the temperature of already frozen products. The temperature fluctuations during transport allow formation of large ice-crystals so the vehicles should be precooled to at least −12°C before loading. The capacity of the vehicle is also important during transferring frozen products. The vehicles having large capacity may take time to be loaded leading to increase heat. Therefore, the usage of the vehicles having large capacity is avoided. The rate of unloading of frozen fish from a vessel depends on many factors, such as the size of blocks or packages, the number and size of the hatches, the degree of accessibility on the vessel, and the number and the skill of the crew that can be employed at one time (Johnston *et al.* 1994). The presence of microbiological residues and odours in the vehicle from previous transfers affects the quality of the frozen materials. The frozen products are prone to damage by improper handling; for example, the tails can be easily broken.

The fish needs to be transferred to low temperatures as soon as possible after removing from the freezer in order to keep its odour, flavour and taste. Keeping at low temperatures decreases the drip-loss during thawing.

8.5 Combination of freezing with traditional and advanced preserving technologies

New technologies are of interest to improve the quality of frozen seafood. High hydrostatic pressure freezing is a novel emerging technology in food processing. Pressure-shift freezing and pressure treatment prior to freezing extends the shelf life of the product and allows to remain the sensorial properties. High hydrostatic pressure extends the shelf life of a product by inactivating microorganisms, and damaging endogenous enzymes, such as lipases, phospholipases, peroxidases and lipoxygenases (Murchie *et al.* 2005). Enzymatic degradation of phospholipids was inhibited during storage at −2 °C for 6 days in cod muscle after the application of pressures above 400 MPa for 15 and

30 min (Oshima *et al.* 1992). High hydrostatic pressure treatment (150, 300, 450 MPa with holding times of 0.0, 2.5, and 5.0 min) prior to freezing to Atlantic mackerel resulted in marked inhibition of free fatty acids and tertiary lipid oxidation compound formation during frozen storage (Vazquez *et al.* 2013). The application of high hydrostatic pressure pretreatments before freezing of Atlantic mackerel changed the activity of acid phosphatase, cathepsins (B and D) and lipase involved in textural deterioration of fish muscle (Fidalgo *et al.* 2014). The pre-treatment of high hydrostatic pressure before freezing and frozen storage was reported to improve some functional and sensory properties in horse mackerel muscle (Torres *et al.* 2014).

Some natural compounds such as chitosan, whey protein isolate, lysozyme, vitamin C, fish gelatin and fish skin hydrolysates are added to biodegradable packaging materials in order to achieve antimicrobial and antioxidant activities. Moreover, the coatings act as a barrier against moisture transfer and oxygen uptake. The protective effect of fish gelatin-based film on horse mackerel patties during frozen storage was observed (Gimenez *et al.* 2011). The whey protein concentrate coating application of Atlantic salmon after freezing increased the thaw yield, decreased the drip-loss, and modified colour parameters of frozen and thawed fillets. Also, it was reported that the protein coatings delayed lipid oxidation of salmon fillets, providing better protection against water glazing (Rodriguez-Turienzo *et al.* 2011). Similar results were obtained during the frozen storage of pink salmon treated with skin hydrolysates, chitosan and protein coatings (Sathivel 2005; Sathivel *et al.* 2008). The usage of natural antioxidant active packaging film derived from barley husks slow down lipid hydrolysis and increase oxidative stability of frozen blue shark (Pereira de Abreu *et al.* 2011).

Glazing of seafood products prevents the incidence of surface drying and dehydration, which may lead to freezer-burn, and quality loss owing to oxidation or rancidity. For this purpose polysaccharides and proteins are used as glazing materials. Chitosan was used as a glazing material for skinless pink salmon, and chitosan nanoparticles were also evaluated for cryogenically frozen shrimp (Sathivel *et al.* 2007; Solval *et al.* 2014). Antioxidants or phenolic substances act as free radical acceptors and not only terminate oxidation at the initial stage but also prevent further formation of new radicals in the oxidation process (O'Sullivan *et al.* 2005). Green tea, grape seed extracts, sage, thyme and clove oils were evaluated as glazing materials for different kinds of marine-derived organisms such as bonito, rainbow trout and shrimp (Sundararajan 2010; Yerlikaya & Gökoğlu 2010; Coban 2013).

Modified atmosphere packaging is another combined preserving technology with freezing. Partially frozen *Venerupis variegate* was packed in modified atmosphere and high density of CO_2 gas found to have better effect on the freshness than that in the samples in ordinary packaging (Fengsheng *et al.* 2013). Packaging in modified atmosphere of cold water shrimp (*Pandalus borealis*) resulted in good quality in relation to colour fading, development of rancid flavour and toughening of the meat (Bak *et al.* 1999).

References

Adams, A.M., Murrell, K.D. & Cross, J.H. (1997). Parasites of fish and risks to public health. *Scientific and Technical Review of the Office International des Epizooties*, **16** (2): 652–660.

Alizadeh, E., Chapleau, N., de Lamballerie, M. & Le-Bail, A. (2007). Effect of different freezing processes on the microstructure of Atlantic salmon (*Salmo salar*) fillets. *Innovative Food Science and Emerging Technologies*, **8**: 493–499.

Aubourg, S. (1998). Influence of formaldehyde in the formation of fluorescence related to fish deterioration. *Zeitschrift für Lebensmittel-Untersuchung und Forschung*, **206**: 29–32.

Aubourg, S.P. (1999). Lipid damage detection during frozen storage of an underutilized fish species. *Food Research International*, **32**: 497–502.

Aubourg, S.P. & Medina, I. (1999). Influence of storage time and temperature on lipid deterioriation during cod (*Gadus morhua*) and haddock (*Melanogrammus aeglefinus*) frozen storage. *Journal of the Science of Food and Agriculture*, **79**: 1943–1948.

Aubourg, S., Lago, H., Sayar, N. & González, R. (2007). Lipid damage during frozen storage of Gadiform species captured in different seasons. *European Journal of Lipid Science and Technology*, **109**: 608–616.

Badii, F. & Howell, N.K. (2002). Texture and structure of cod and haddock fillets during frozen storage. *Food Hydrocolloids*, **16**: 313–319.

Bak, L.S., Andersen, A.B., Andersen, E.M. & Bertelsen, G. (1999). Effect of modified atmosphere packaging on oxidative changes in frozen stored cold water shrimp (*Pandalus borealis*). *Food Chemistry*, **64**: 169–175.

Barthet, V. J., Gordon, V. & Dany, J. (2008). Evaluation of a colorimetric method for measuring the content of FFA in marine and vegetable oils. *Food Chemistry*, **111**: 1064–1068.

Benjakul, S., Visessanguan, W., Ishizaki, S. & Tanaka, M. (2000). Differences in gelation characteristics of natural actomyosin from two species of bigeye snapper, *Priacanthus tayenus* and *Priacanthus macracanthus*. *Journal of Food Science*, **66**: 1311–1318.

Benjakul, S., Visessanguan, B., Thongkaewa, C. & Tanaka, M. (2003). Comparative study on physicochemical changes of muscle proteins from some tropical fish during frozen storage. *Food Research International*, **36**: 787–795.

Benjakul, S., Visessanguan, W., Thongkaew, C. & Tanaka, M. (2005). Effect of frozen storage on chemical and gel-forming properties of fish commonly used for surimi production in Thailand. *Food Hydrocolloids*, **19**: 197–207.

Boyer, R. & McKinney, J. (2013). Food storage guidelines for consumers. *Virginia Cooperative Extension Publication*, **348–960**.

Brady, Y.J. & Vinitnantharat, S. (1990). Communications: Viability of bacterial pathogens in frozen fish. *Journal of Aquatic Animal Health*, **2** (2): 149–150.

Castro-Escarpulli, G., Figueras, M.J., Aguilera-Arreola, G., *et al.* (2003). Characterisation of Aeromonas spp. isolated from frozen fish intended for human consumption in Mexico. International *Journal of Food Microbiology*, **84**: 41–49.

Cemeroğlu, B. (2005). Gıda Mühendisliğinde Temel İşlemler. Gıda Teknolojisi Derneği Yayınları, No:29, 505 pp.

Chevalier, D., Sequeira-Munoz, A., Le Bail, A., Simpson, B.K. & Ghoul, M. (2001). Effect of freezing conditions and storage on ice-crystal and drip volume in turbot (*Scophthalmus maximus*): Evaluation of pressure shift freezing vs. airblast freezing. *Innovative Food Science and Emerging Technologies*, **1** (3): 193–201.

Coban, O.E. (2013). Evaluation of essential oils as a glazing material for frozen rainbow trout (*Oncorhynchus mykiss*) filet. *Journal of Food Processing and Preservation*, **37**: 759–765.

Diaz-Tenorio, L., Garcia-Carreno, F.L. & Pacheco-Aguilar, R. (2007). Comparison of freezing and thawing treatments on muscle properties of whiteleg shrimp (*Litopenaeus vannamei*). *Journal of Food Biochemistry*, **31**: 563–576.

Dobarganes, C. & Márquez-Ruiz, G. (2003). Oxidized fats in foods. *Current Opinion in Clinical Nutrition and Metabolic Care*, **6** (2): 157–163.

Duan,J., Cherian, G. & Zhao, Y. (2010). Quality enhancement in fresh and frozen lingcod (*Ophiodon elongates*) fillets by employment of fish oil incorporated chitosan coatings. *Food Chemistry*, **119**: 524–532.

Einen, O., Guerin, T., Fjæra, S.O. & Skjervold, P.O. (2002). Freezing of pre-rigor fillets of Atlantic salmon. *Aquaculture*, **212**: 129–140.

Erkan, N., Varlık, C. 2004. Dondurarak muhafaza teknolojisi. Su Ürünleri İşleme Teknolojisi, pp. 96–127. Istanbul Universitesi Yayin no: 4465, Istanbul.

Fengsheng, Q., Hongying, L. & Haojie, Z. (2013). Effects of modified atmosphere packaging on fresh preservation of partially frozen *Venerupis variegata*. *Journal of Biobased Materials and Bioenergy*, **7** (2): 290–294.

Fidalgo, L., Saraiva, J.A., Aubourg, S.P., Vazquez, M. & Torres, J.A. (2014). Effect of high pressure pre-treatments on enzymatic activities of Atlantic mackerel (*Scomber scombrus*) during frozen storage. *Innovative Food Science and Emerging Technologies*, **23**: 18–24.

Garthwaite, G.A. (1997). Chilling and freezing of fish. In: *Fish Processing Technology* (Ed. Hall, G.M.), pp. 93–117, Chapman & Hall, London.

Gill, C.O. (2006). Microbiology of frozen foods. In: *Handbook of Frozen Food Processing and Packaging* (Ed. Sun, W), pp. 85–95, CRC Press, Taylor & Francis Group, New York.

Gimenez, B., Gomez-Guillen, M.C., Perez-Mateos, M., Montero, P. & Marquez-Ruiz, G. (2011). Evaluation of lipid oxidation in horse mackerel patties covered with borage-containing film during frozen storage. *Food Chemistry*, **124**: 1393–140.

Gökoğlu, N. (2002). *Su Ürünleri İşleme Teknolojisi*. pp. 157, Su VakfıYayınları, Istanbul.

Hui, Y.H., Legaretta, I. G., Lim, M.H., Murrell, K.D. & Nip, W. (2004). *Handbook of Frozen Foods*, pp. 1293. CRC Press, Boca Raton.

Hwang, K. T. & Regenstein, J. M. (1989). Protection of menhaden mince lipids from rancidity during frozen storage. *Journal of Food Science*, **54**: 1120–1124.

Jeremiah, L. E. (1996). *Freezing Effects on Food Quality*, pp. 432, CRC Press, Marcel Dekker, Inc. New York.

Johnston, W.A., Nicholson, F.J., Roger, A. & Stroud, G.D. (1994). Freezing and refrigerated storage in fisheries, pp. 143. FAO Fisheries Technical Paper 340, Food and Agriculture Organization, Rome, Italy.

Kolbe, E. & Kramer, D. (2007). Planning for seafood freezing, pp. 112. Alaska Sea Grant College Program, Cooper Publishing, Alaska.

Lund, M. N., Heinonen, M., Baron, C. P. & Estévez, M. (2011). Protein oxidation in muscle foods: a review. *Molecular Nutrition & Food Research*, **55**: 83–95.

Mazur, P. (1966). Physical and chemical basis of injury in single-celled microorganisms subjected to freezing and thawing. In: *Cryobiology* (Ed. Meryman, H.T.), pp. 213–315, Academic Press, London.

Mackie, I. M. (1993). The effect of freezing on fish proteins. *Food Review International*, **9** (4): 575–610.

Murchie, L., Cruz-Romero,M., Kerry, J., *et al.* (2005). High pressure processing of shellfish: A review of microbiological and other quality aspects. *Innovative Food Science and Emerging Technologies*, **6**: 257–270.

Ohshima, T., Nakagawa, T., & Koizumi, C. (1992). Effect of high hydrostatic pressure on the enzymatic degradation of phospholipids in fish muscle during storage. In: *Seafood Science and Technology*, Chapter 8 (Ed. Bligh, E.), pp. 64–75, Wiley-Blackwell, Oxford, UK.

O'Sullivan, A., Mayr, A., Shaw, N. B., Murphy, S. C. & Kerry, J. P. (2005). Use of natural antioxidants to stabilize fish oil systems. *Journal of Aquatic Food Product Technology*, **14**: 75–94.

Pan , B.S. & Chow, C.J. (2004). Freezing secondary seafood products. In: *Handbook of Frozen Foods* (Eds. Nollet, L.M.L. & Toldra, F.), pp. 325–339, CRC Press, Boca Raton.

Pereira de Abreu, D.A., Losada, P.P., Maroto, J. & Cruz, J.M. (2010). Evaluation of the effectiveness of a new active packaging film containing natural antioxidants (from barley husks) that retard lipid damage in frozen Atlantic salmon (*Salmo salar* L.). *Food Research International*, **43**: 1277–1282.

Pereira de Abreu, D.A., Losada, P.P., Maroto, J. & Cruz, J.M. (2011). Natural antioxidant active packaging film and its effect on lipid damage in frozen blue shark (*Prionace glauca*). *Innovative Food Science and Emerging Technologies*, **12**: 50–55.

Regost, C., Jakobsen, J.V. & Rora, A.M.B. (2004). Flesh quality of raw and smoked fillets of Atlantic salmon as influenced by dietary oil sources and frozen storage. *Food Research International*, **37**: 259–271.

Rodriguez-Turienzo, L., Cobos, A., Moreno, V., Caride, A., Vieites, J.M. & Diaz, O. (2011). Whey protein-based coatings on frozen Atlantic salmon (*Salmo salar*): influence of the plasticiser and the moment of coating on quality preservation. *Food Chemistry*, **128**: 187–194.

Saeed, S. & Howell, N. K. (2002). Effect of lipid oxidation and frozen storage on muscle proteins of Atlantic mackerel (*Scomber scombrus*). *Journal of the Science of Food and Agriculture*, **82**: 579–586.

Sánchez-Alonso, I., Martinez, I., Sánchez-Valencia, J. & Careche, M. (2012). Estimation of freezing storage time and quality changes in hake (*Merluccius merluccius*, L.) by low field NMR. *Food Chemistry*, **135**: 1626–1634.

Sathivel, S. (2005). Chitosan and protein coatings affect yield, moisture loss, and lipid oxidation of pink salmon (*Oncorhynchus gorbuscha*) fillets during frozen storage. *Journal of Food Science*, **70**: 455–459.

Sathivel, S., Liu, Q., Huang, J. & Prinyawiwatkul, W. (2007). The influence of chitosan glazing on the quality of skinless pink salmon (*Oncorhynchus gorbuscha*) fillets during frozen storage. *Journal of Food Engineering*, **83**: 366–373.

Sathivel, S., Huang, J. & Bechtel, P. J. (2008). Properties of pollock (*Theragra chalcogramma*) skin hydrolysates and effects on lipid oxidation of skinless pink salmon (*Oncorhynchus gorbuscha*) fillets during 4 months of frozen storage. *Journal of Food Biochemistry*, **32**; 247–263.

Siddaiah, D., Reddy, G.V.S., Raju, C.V. & Chandrasekhar, T.C. (2001). Changes in lipids, proteins and kamaboko forming ability of silver carp (*Hypophthalmichthys molitrix*) mince during frozen storage. *Food Research International*, **34**: 47–53.

Sigurgisladottir, S., Ingvarsdottir, H., Torrissen, O.J., Cardinal, M. & Hafsteinsson, H. (2000). Effects of freezing/thawing on the microstructure and the texture of smoked Atlantic salmon (*Salmo salar*). *Food Research International*, **33**: 857–865.

Solval, K.M., Rodezno, L.A.E., Moncada, M., Bankston, J.D. & Sathivel, S. (2014). Evaluation of chitosan nanoparticles as a glazing material for cryogenically frozen shrimp. *LWT - Food Science and Technology*, **57**: 172–180.

Sørensen, N.K., Brataas, R., Nyvold, T.E. & Lauritzen, K. (1997). Influence of early processing (pre-rigor) on fish quality. In: *Seafoods from Producer to Consumer, Integrated Approach to Quality* (Eds Luten, J.B., Børressen, T. & Oehlenschlager, J.), pp. 253– 263, Elsevier, Amsterdam.

Sundararajan, S. (2010). Evaluation of green tea extract as a glazing material for shrimp frozen by cryogenic and air-blast freezing. pp. 84, Master of Science Thesis. B.Tech., Vellore Institute of Technology University.

Sun, D. (2006). *Handbook of Frozen Food Processing and Packaging*, pp. 732, CRC Press Taylor & Francis Group, New York.

Tejada, M., Mohamed, G.F., Huidobro, A. & Garcia, M.L. (2003). Effect of frozen storage of hake, sardine and mixed minces on natural actomyosin extracted in salt solutions. *Journal of the Science of Food and Agriculture*, **83**: 1380–1388.

Torres, J.A., Saraiva, J.A., Guerra-Rodríguez, E., Aubourg, S.P. & Vázquez, M. (2014). Effect of combining high-pressure processing and frozen storage on the functional and sensory properties of horse mackerel (*Trachurus trachurus*). *Innovative Food Science and Emerging Technologies*, **21**: 2–11.

Vazquez, M., Torres, J.A. Gallardo, J.M., Saraiva, J. & Aubourg, S.P. (2013). Lipid hydrolysis and oxidation development in frozen mackerel (*Scomber scombrus*): Effect of a high hydrostatic pressure pre-treatment. *Innovative Food Science and Emerging Technologies*, **18**: 24–30.

Venugopal, V. (2006). Quick freezing and individually quick frozen products. In: *Seafood Processing*, pp. 446, Taylor & Francis Group, CRC Press, New York.

Xiong, Y.L. (1997). Protein denaturation and functionality losses. In: *Quality in Frozen Food* (Eds Erickson, M. & Hung, Y.-C.), pp. 111–140, Chapman & Hall/ International Thomson Publishing, New York.

Yerlikaya, P. & Gökoğlu, N. (2010). Inhibition effects of green tea and grape seed extracts on lipid oxidation in bonito fillets during frozen storage. *International Journal of Food Science and Technology*, **45**: 252–257.

CHAPTER 9

Thawing of fish

9.1 Quality changes of fish during thawing

The water content of food crystallizes during freezing, whereas these ice crystals liquefy in the thawing process. Thawing is physically the reverse of freezing. The surface of the frozen fish thaws first when exposed to heat. The remaining frozen fish is surrounded by thawed material with about one-third the thermal conductivity of the frozen material; consequently the time necessary to thaw fish is much greater than that necessary to freeze it under similar conditions of heat transfer (Bezanson *et al.* 1973). The time required for the thawing process is longer than that for freezing, depending on two factors. First, the temperature difference between the frozen food and the thawing medium is small. Second, the difference between thawing and freezing depends on the physical characteristics of water and ice, such as coefficient of thermal conductivity and thermal diffusion coefficient (Table 9.1). The temperature of ice changes nine times faster than that of water (Cemeroğlu 2005).

A layer of ice forms on the surface of the freezing material during the freezing process. Underneath that covering, there is another ice layer, generating latent heat of freezing (335 kJ/kg). Latent heat is

Seafood Chilling, Refrigeration and Freezing: Science and Technology, First Edition.
Nalan Gökoğlu and Pınar Yerlikaya.
© 2015 John Wiley & Sons, Ltd. Published 2015 by John Wiley & Sons, Ltd.

Table 9.1 Physical characteristics of water and ice

Physical characteristics	Water	Ice
Coefficient of thermal conductivity at 0°C	0.561 W/mK	2.24 W/mK
Thermal diffusion coefficient at 0°C	1.3×10^7 m^2/s	11.7×10^7 m^2/s

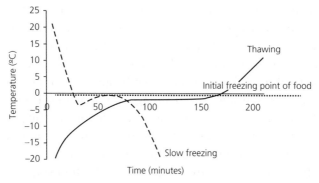

Figure 9.1 Temperature changes of food with respect to time during freezing and thawing.

associated with the change of phase and this heat can be easily transferred out quickly from the upper layer due to the high coefficient of thermal conductivity and diffusion through the ice. However, during thawing, the outer layer of the frozen material reaches the thawing temperatures and the ice turns into water. Water acts like an insulator and slows the thawing process. The thawing rate should be slow enough to allow the water to change its original position to avoid drip-loss. Therefore thawing should be as quick as freezing, in order to shorten the time of exposure to high temperatures. It is known that microbiological and enzymatic reactions are accelerated at high temperatures, so rapid thawing at low temperatures is the best way of keeping the quality of frozen product. Figure 9.1 shows the temperature changes of food during freezing and thawing.

A large amount of heat energy is required to disrupt the hydrogen-bonded lattice of ice. Each water molecule participates in four hydrogen bonds in the common form of ice. When ice melts, most of the hydrogen bonds are retained irregularly in water, due to fluctuation. The average energy required to break each hydrogen bond in ice has been estimated to be 23 kJ/mol, whereas this value is less than 20 kJ/mol in water. A relatively large amount of heat is required to raise the temperature of 1 g of water by 1°C. The unusual thermal properties of water make it a suitable environment for living organisms and chemical reactions (Cybulska & Doe 2002).

The freezing process proceeds from the exterior waters on the surface to the intracellular water. Ice will crystallize and the solutes of the tissue are concentrated. These enzymes and other molecules tend to react rapidly between −2°C and −5°C, which is called the critical zone and needs to be passed quickly. This zone is also the optimum temperature for denaturation of proteins. Satisfactory results in the final quality of the fish can be achieved by good initial quality, as well as conditions of freezing, frozen storage and thawing. The cold chain and hygienic conditions also need to be considered during processing.

The changes during freezing and frozen storage continue in the thawing process. The majority of thawing time takes place around, or slightly below, the freezing point. Microbiological activity, recrystallization and chemical reactions take place rapidly in this zone. The thawing process itself is a factor that damages foods through chemical and physical changes and by contamination by microorganisms. Rapid thawing at low temperatures helps to prevent the loss of fish quality during the thawing process. Once the seafood is thawed, it will spoil as quickly as chilled or fresh seafood. Therefore, thawed food must be kept chilled or processed as soon as possible.

Water is removed from its original location and forms ice crystals within the food during the freezing process. The size and location of the ice crystals are required to be as small as possible where they are present. Large intracellular crystals cause maximal damage to the cell wall. Fluctuating temperatures during frozen storage and long storage times encouraging recrystallization lead to tissue damage. When the fish is thawed, the water is no longer bound to the muscle proteins and drains away as drip-loss. The tissue mostly does not have the ability to reabsorb the melted ice crystals. Drip-loss may be up to 5% of the fresh fish weight, under the conditions of a proper freezing process. The factors affecting drip-loss are the form of ice crystals and their location, rate of thawing, water reabsorption, the physiological and biochemical status of the tissue prior to freezing, and the intrinsic water binding strength in the tissue (Erickson &

Hung 1997). High drip-losses are undesirable due to being visually unattractive, causing loss of water-soluble nutrients, flavour compounds and weight, resulting in a dry and stringy texture (Kolbe & Kramer 2007).

When muscle is frozen at the pre-rigor stage and kept in cold storage, it is still able to contract and go into rigor after thawing. This is known as thaw rigor. The process of rigor must be completed slowly during frozen storage so that rigor does not occur during thawing. When pre-rigor fillets are thawed, the muscle will contract as soon as the ice within the flesh has melted, with a loss of cellular fluid and much thaw drip. Thaw rigor results in tough muscle tissue and a lot of gaping (breaking of connective tissue between the muscle segments). To avoid thaw drip, rubbery texture, and serious shrinkage, fish or fillets frozen solid before the onset of rigor mortis must be thawed slowly at a low temperature. The energy source for muscle contraction during thawing is ATP. Thaw rigor can be avoided by ending ATP synthesis and degradation at subfreezing temperatures while the muscle is still frozen. Another approach to avoid thaw rigor is by conditioning, which is when the fish is held at a relatively high temperature in the freezing process to allow completion of rigor while the muscle is still frozen (Kolbe & Kramer 2007).

A part of the cellular fluid is released after thawing, leading to microbial growth. As mentioned before, the thawing process takes place around or slightly below the freezing point. The surrounding medium temperature is required to be less than 15°C during thawing. These temperatures are required by psychrotrophic microorganisms. Once thawed, the microorganisms that have survived the freezing and frozen storage will grow and multiply under conditions that can lead to foodborne illness. Thawed seafood is a good substrate for microbiological growth, having a moist surface with respect to drip-loss and condensation. Minimal ambient temperature must be ensured. Microbiological flora, initial fish quality, freezing process, frozen storage temperature and fluctuations, thawing method and storage conditions, all affect the microbiological quality of the thawed product.

During thawing, the growing ice-crystals create pressure on microbial cells, as observed during the freezing process. Also, the medium surrounding the microbial cells is diluted, allowing the cells to be exposed to osmotic shock and leading to inhibition of microbiological activity. It is appreciated that increasing the number of freeze–thaw cycles leads to greater loss of microbiological viability. However, the use of repeated freeze–thaw cycles is not acceptable because of the textural properties and other consumer acceptance criteria (Archer et al. 2008).

One of the major causes of the changes during thawing is the reactions catalysed by enzymes. The enzymes remains stable while in the frozen state and may be reactivated very rapidly after thawing. Freezing and thawing disrupt muscle cells, resulting in the release of enzymes from mitochondria into the sarcoplasm (Sriket et al. 2007). Proteasomes, matrix metalloproteinases, calpains and lysosomal cathepsins are the proteolytic systems present in fish muscle tissue playing major roles in the degradation of fish muscle proteins (Delbarre-Ladrat et al. 2006). The enzymatic reactions accelerate in this period due to the presence of the intracellular material and release of enzymes. Slow freezing produces larger extracellular ice-crystals resulting in more tissue damage. This damage involves a release of proteases (calpains and cathepsins) which are able to hydrolyse myofibrillar proteins and then lead to textural changes (Doughikollaee 2012).

Nutrient loss occurs in the frozen product due to drip-loss during the thawing process. The amount of water-soluble proteins, vitamins and minerals leaches with the loss of water. The freeze–thaw process promotes protein oxidation and protein denaturation, which is associated with the formation of disulphide bonds. Also, lipid oxidation occurs during the freeze–thaw process, leading to nutritional and textural changes (Srinavasan et al. 1997). The lipid oxidation products such as hydroperoxides cause oxidative alterations in sulphur-containing proteins.

Generally, frozen food material loses its initial structure. One of the major deteriorative changes in the fish muscle is myofibrillar protein denaturation, which can lead to textural and functional changes. The drip-loss increases dramatically with protein denaturation during freezing and frozen storage. The denaturation of the protein is followed by aggregation, which involves cross-links between protein molecules. There is a direct correlation between drip-loss and texture. Excessive drip-loss during thawing will result in a dry, stringy texture (Kolbe & Kramer 2007) and the thawed material becomes drier, tougher and less tasty. Textural changes during thawing depend on the shape of frozen fish, even if it is block frozen. Block fish has less deformation compared to filleted fish because of the structural support of the fishbone. Thawed meat tends to display higher shear force than does unfrozen meat (Hale & Waters 1981).

Colour changes may occur in frozen fish or crustaceans, depending on the thawing method. Melanosis takes place in shrimps that are thawed over 0°C. A decrease in the intensity of initial fresh odour and flavour is observed in frozen seafood. The sense of rancid odour and flavour is obvious especially in fatty fish because of lipid deterioration. Protein oxidation during freezing and thawing leads to a decrease in sensory qualities such as flavour, tenderness and juiciness.

9.2 Thawing methods of frozen fish

The important points of thawing are avoiding overheating the frozen material, protecting excessive drip-loss, protection of microbiological growth, and performing the thawing in a limited time. The method of thawing depends on the frozen material and freezing technique. Optimum thawing procedures should be of concern to food technologists. The appropriate thawing process is required to obtain a product similar to the fresh one.

The thawing method can be distinguished by way of heat supply, such as heating the surface of the product and transfer of the heat by conduction, and by generating the heat within the product (Keizer 1995). In the first group of methods, air, water or brine is used as a heating source and the outside of the fish thaws first. In the second group, heat forms in the interior of a product by application of electric current or electric field, radio waves or microwaves.

9.2.1 Thawing with air

Thawing with air means exposing the frozen material to warm air at about 15–21°C. The frozen fish leaves on a plate and is heated by convection. There is a weak heat transfer between the surface of the material and the air. Surface heat transfer depends on the relative humidity of the air, so air saturated with water vapour is preferred. In this case, a high relative humidity prevents the product from dehydrating. Packaging of the material is another method of protection from dehydration. However, packaging the product extends the thawing time.

Frozen fish, fish fillets or fish blocks can be thawed by leaving an overnight at room temperature. Still air thawing is slower than thawing in moving air. Thawing with air is mostly preferred at home, but is also used in industry under controlled conditions. Blast or high air flow is used in industrial applications. The air flow speed is around 8–10 m/s in order to obtain uniform thawing. In this case, the film coefficient between the surface of the material and the air decreases, leading to an increase in heat transfer. Air-blast thawing needs a permanent set-up. The air flows parallel to or opposite to the fish that is lying on the circulation band. The flow should be uniform and the temperature of the air should not exceed 20°C, otherwise, lipid oxidation, recrystallization and texture chances take place.

Time of thawing depends on many factors, such as specific heat of the product, speed of air flow, and the ratio of surface area to the volume of the product. The time of thawing will be faster in greater surface areas of the fish exposed. Slow thawing taking days should be avoided because the temperature of the outer layers of frozen fish will be suitable for microorganisms to grow before the centre is thawed.

It is important to avoid dripping of thawing water from one fish to another on the shelves. This dripping may cause bacteriological contamination, colour disorders and insufficient thawing.

9.2.2 Thawing with water

Thawing with water is performed by dipping the frozen material into still or flowing water, or by spraying the water onto the frozen material. This method is simple, economic and suitable for short-time and continuous batches. The temperature of the water should be less than 18°C and the circulation speed of water should be 5 mm/s (Gökoğlu 2002). The speed of water flow is important in the thawing process, as in the method of thawing with air. The thawing rate increases with the increasing speed of water flow.

Tanks are used in this thawing method. The rate of fish to water is 1/4 or 1/5. This thawing method is proper for small or medium-sized fish. Big fish should be held for long periods, which leads to swelling and quality losses. Filleted fish absorb high amounts of water and losses may occur in the appearance and flavour of the product. Continuous thawing systems can be applied but scales and small pieces of the fish is a serious problem in continuous immersion thawing. Moreover, the circulated water needs to be cleaned. The circulation should be sufficient to produce even thawing.

The surface of the heat transfer coefficient is higher in water thawing than air thawing. This method has the advantage of no need for mechanization and no problem with dehydration; moreover, it gives weight gain to the product. However, the product that is held in the water can swell; unpackaged products can lose flavour and odour, and contaminants can pollute the water.

Different thawing treatments, such as thawing in a refrigerator, in water, in air at ambient temperature and in a microwave oven were compared for frozen eel, and the lowest total aerobic mesophilic bacteria count and yeast count were determined in water-thawed samples by Ersoy *et al.* (2008).

9.2.3 Thawing under vacuum

Fish that are placed in a closed system are exposed to a vacuum. Low-pressure steam leaves its latent heat of evaporation while the vapour condenses on the cold surface of the frozen fish in the absence of air. Thus, the lipid oxidation of fish is avoided.

The system of thawing under vacuum is composed of a vacuum cell, a vapour source, a vacuum pump and the necessary controls. The seafood is loaded onto trolleys into the airtight chambers. A vacuum is drawn in the chamber and water in a reservoir within the base of chamber is heated to produce water vapour. The ratio of applied vacuum depends on the thawing temperature. Vapour is injected at constant temperature and condensation occurs on the cold surface of the fish, allowing the heat released to be transferred rapidly and absorbed by the material. Thawing process takes place in short time; however, it has to be programmed for each product.

9.2.4 Thawing with electrical resistance

An alternating electrical current passes through the food sample, resulting in internal energy generation in the frozen food. The food material acts as a resistor in an electrical circuit. The electrical resistance of the food translated into heat. The heat is generated throughout the entire mass of the food uniformly (Figure 9.2). If an electrical current is passed through a conductive material, the temperature of the material increases, depending on the electrical conductivity. The frozen

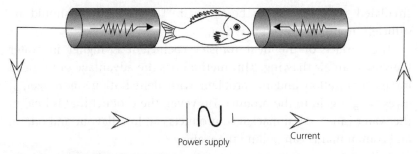

Power supply   Current

Figure 9.2 Thawing with electrical resistance.

product is brought into good contact with electrodes imposing an alternating electrical current to the product (Keizer 1995). The heat of the product increases rapidly after placing the frozen fish between two stainless steel plaques where an electrical current is applied. Electrical resistance of the fish decreases with the increasing heat, so electrical current passes through the warmer fish blocks leading to heating of the product.

In practice, fillets of small whole fish in blocks up to 50 mm thick are first immersed in cold water for 15–30 minutes in order to conduct the electrical current. Then, the fish are heated electrically for a further 15–20 minutes until thawed (Jason 2001).

Thawing with electrical resistance is two to three times faster than thawing with air and water. There is no limit for heat penetration, as in microwave thawing. This method helps to maintain the physiochemical aspects and nutrient content. No localized overheating is observed. This method is also called ohmic thawing.

It is reported that the voltage gradient applied during ohmic thawing was shown to be a vital factor influencing the textural properties of beef cuts in terms of the hardness, springiness, gumminess and chewiness (Yildiz-Turp *et al.* 2013). An ohmic thawing unit was designed for shrimp blocks comprising computer-automated surface temperature sensors without runaway heating problems(Roberts *et al.* 1998). The frozen shrimp blocks and fish were thawed by passing an alternating current through the blocks (Balaban *et al.* 1994). The electrical conductivities of frozen shrimp and flounder were also reported by Luzuriaga and Balaban (1996). The amount of protein destruction and the removal of fat globules were found to be higher in conventionally thawed samples than ohmic thawed samples (Icier *et al.* 2010).

9.2.5 Dielectric thawing

Dielectric energy is a form of electromagnetic energy. The waves are transmitted to food, absorbed, and converted to heat within the product, as with ohmic heating. The methods differ in frequency of electrical energy used. The heating is applied at lower frequencies (10–80 MHz) alternating voltage (~5 kV) in dielectric thawing, at moderate frequencies (1–10 GHz) in microwave thawing and at high frequencies (50 Hz and 220 V) in electrical resistance thawing (Haugland 2002).

The heating is obtained by the internal friction of molecules due to oscillation of dipoles and movements of ions in the frequently changing electric fields. The increase in the temperature of water molecules heats the surrounding components within the food material. Frozen blocks are placed between two metal plates, which are contacted with an electric source of high voltage and frequency. The alternative electrical field formed between the plaques causes heat inside the product, regardless of the contact characteristics between the plates and the product. The heat that is formed by the movement of charged particles in the product against the alternative field is absorbed by the product, which thaws rapidly. A dielectric defrosting process was applied to fish blocks by using electrodes at 35 MHz, and it was observed that there was a marked reduction in thawing time from 20–30 hours in conventional cold room defrosting to 20–30 minutes (Bengtsson 1963). If the frozen fish blocks are immersed into water before dielectric thawing, the electrical conditions of the within the fish will be uniform and the performance of the thawing will be improved.

Dielectric thawing can be mechanized easily and continuity can be achieved. The products are placed on bands carrying them between the plaques. The proper thawing can be achieved by adjusting the speed of the conveyor, size of the plaques, voltage and frequency. Local overheating can be observed in case of irregular shape of fish blocks or there being warmer areas of the frozen fish before the dielectric thawing process has been applied.

9.2.6 Microwave thawing

Microwaves are electromagnetic waves with longer wavelengths than radio waves and infrared radiation. They are measured in metres or centimetres and the frequency band of microwave is 2450 MHz in general. Food absorbs microwaves and converts them to heat. Heating of a food

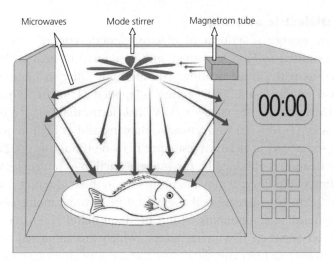

Microwaves   Mode stirrer   Magnetrom tube

00:00

Figure 9.3 Microwave thawing of fish.

with microwave energy depends on the 'dipole' molecules contained in it. Water is a characteristic dipole molecule, with its positive hydrogen ions and negative oxygen ions. Dipole molecules change direction billions of times per second, depending on frequency, in order to accommodate the electric field. An intense friction occurs, leading to heating of the food material (Cemeroglu 2005). Heat is transferred, molecule by molecule, from the outside of the frozen material in convectional heating. This enables there to be a heat gradient between the centre and the outer edge of the frozen material. Microwave thawing needs no direct contact with the food (Figure 9.3). In microwave heating, dipole molecules react attempting to align in the material and generate heat which is distributed throughout the material. The thawing process occurs on each side of the fish due to containing high amount of water.

The thawing rate depends on material properties and dimensions, the magnitude and frequency of the electromagnetic radiation (Pangrle *et al.* 1991). High salt content increases the uniformity of the frozen material, allowing a decrease in penetration depth and accelerating the energy of microwaves onto the surface of the material. The microwave permeability of water is greater than of ice. The rate of heating of a product increases after thawing. Thawing with microwave is 10 times faster than dielectric thawing, needs smaller space for processing and reduces drip-loss; however, it is not useful in industrial applications.

In a study of some quality characteristics of frozen rainbow trout based on different thawing methods, it was found that thawing in the refrigerator and under running water were the most suitable thawing methods to keep the physical quality of the fish. Thawing in a microwave oven is not recommended because of causing poor quality in terms of the physical attributes of fish (Yerlikaya & Gökoğlu 2005).

Headless brown shrimps were thawed using microwaves in a 915 mHz conveyorized multimode applicator. It was found that this process was suitable for thawing of shrimps because:

1 Microwave defrosting would allow compliance with the present good manufacturing practice (GMP) guideline for raw, headless shrimp regarding the requirements of temperature and packaging removal.
2 There is improved production control, resulting from rapid in-line processing.
3 Water usage is reduced substantially, alleviating waste disposal problems.
4 Defrosting takes place within the carton, eliminating the need to remove the carton and increasing handling efficiency after thawing.
5 Ice requirements are reduced because there is no 'temperature overshoot'.
6 Bacteriological control and quality control are improved.
7 Wholesomeness, as evidenced by the moisture/ protein ratio, is retained.
8 Total defrosting cost is reduced compared with air or water defrosting (Bezanson *et al.* 1973).

9.2.7 Thawing with hydrostatic high pressure

The method of high-pressure processing is to transfer the pressure uniformly and rapidly throughout the material, without considering the size or dimensions. The phase transition temperature of ice is lowered under pressure, allowing the frozen sample to be thawed at subzero temperatures. Low temperatures and pressure have synergetic effect on the inhibition of microbiological growth.

It is suggested that rapid thawing at low temperatures helps to maintain the initial quality of the food. However, this phenomenon is a challenge in that application of lower temperatures reduces the temperature differential between the frozen sample and the ambient temperature (Alizadeh *et al.* 2007). The application of high pressure

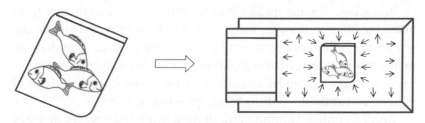

Figure 9.4 High hydrostatic pressure thawing of fish.

allows enlarging the temperature differential between the frozen sample and the surrounding ambient temperature, which is the principal driving force for the thawing process. The thawing time reduces depending on the enhanced heat flux rate. Assuming that thawing is carried out at 20°C, conventional thawing gives a temperature difference of about 20°C, while at 200 MPa the difference will be approximately twice as much. Thus, high-pressure thawing time at 200 MPa can be theoretically expected to be half of conventional immersion thawing time (Zhu *et al.* 2004). The application of high pressure remains the water in a liquid state to a minimum of −22°C at 209–210 MPa (Bridgman 1912; Wagner *et al.* 1994). Figure 9.4 shows the application of high hydrostatic pressure schematically.

It is better to define the type of high pressure thawing process into pressure-assisted thawing and pressure-induced thawing. In the pressure-assisted thawing process the phase transition occurs under constant pressure by increasing the temperature. In the pressure-induced thawing process the phase transition is initiated by a pressure change and is continued at constant pressure (Knorr *et al.* 1998).

Phase change of water is obtained by both heating at constant pressure and pressurization, depending on the initial temperature of the sample, rate of pressurization, working pressure, sample size, and other conditions regarding process control (Schubring *et al.* 2003). The melting point of water is depressed under pressure. Pressure induces the formation of smaller and more regular ice-crystals and hinders their expansion.

The process after thawing under high pressure is also important in that care must be taken to increase the temperature of the sample enough in order not to produce ice crystallization during the expansion

(Otero & Sanz 2003). The temperature of the sample must be brought to a minimum level above 0°C before releasing the pressure to avoid the formation of ice-crystals (Doughikollaee 2012).

The main effect of high pressure is to trigger the changes in hydrophobic and electrostatic interactions leading to alterations of secondary, tertiary and quaternary structures of the proteins. The functionality of the proteins is affected because ofconformational changes (Mozhaev *et al.* 1996).

High-pressure thawing of frozen meat required only one-third of the time necessary at atmospheric pressure (Makita 1992). An improvement of the microstructure in sea bass muscle is achieved by pressure shift freezing, and the pressure-assisted thawing application remains unchanged during frozen storage due to an instantaneous and uniform nucleation throughout the sample, producing small ice-crystals (Tironi *et al.* 2010). The application of high pressure thawing on various fish species such as whiting fillets and tuna meat shortens the thawing time, decrease the drip-loss, and lessen the microbiological load (Murakami *et al.* 1992; Chevalier *et al.* 1999; Schubring *et al.* 2003). Organoleptic characteristics of high-pressure treated various fish fillets were reported to be better compared to the conventional water-thawed samples (Schubring *et al.* 2003).

Limitations on the application of high-pressure thawing are mainly the high cost and pressure-induced protein denaturation and aggregation, discoloration and textural changes (Chevalier *et al.* 2000; Li & Sun, 2002). Colour changes in salmon fillets treated by high-pressure thawing became significant when thawed at pressures above 150 MPa (Zhu *et al.* 2004). The increase in L*, a*, b* values (L* describes lightness, a* redness and b* yellowness) were observed in high-pressure thawing-treated tuna meat (Murakami *et al.* 1992).

9.2.8 Ultrasound-assisted thawing

Ultrasound is sound waves above the limit of the human hearing range of ~20 kHz (Awad *et al.* 2012). The uses of ultrasound in food industry is classified into two groups; The first one is high-frequency, low-energy, diagnostic ultrasound, involving the use of frequencies higher than 100 kHz at intensities below 1 W/cmP²P. Ultrasonic waves cause no physical or chemical alterations in the properties of the material through which the wave passes. They are used for non-invasive monitoring of food processes. The second use is low-frequency, high-energy, power ultrasound in the kHz range. 0Thigh-power ultrasound has mechanical, chemical and/or biochemical effects, which are used to modify the physicochemical properties and enhance the quality of various food systems during processing (Jambrak 2012). 0TUltrasonic waves in this range are capable of altering material properties through generation of immense pressure, shear and temperature gradient in the medium through which they propagate (Dolatowski *et al.* 2007).

Sonication is a tool in controlling the crystallization process, enhancing the nucleation rate and ice-crystal growth rate and allowing generating new and fresh nucleation sites (Bermudez-Aguirre *et al.* 2011). Acoustic thawing is a promising technology in the food industry to shorten thawing time, thus reducing cell damage, drip-loss and improving product quality. Thawing process under the relaxation frequency is faster than that obtained with a thawing process using only conductive heating (Chemat *et al.* 2011; Mason *et al.* 2011). The energy is transferred to the product through vibrations caused by a high-frequency sound source (0.2–4 MHz). Direct contact with the product is not necessary but is preferable. The ultrasound is more highly attenuated in frozen fish than in unfrozen flesh and the attenuation increases markedly with temperatures, reaching a maximum near the initial freezing point of the food. This means that most of the energy should be absorbed at the frozen/thawed boundary (Haugland 2002).

Application of ultrasound in thawing food has been thought to have negative aspects such as poor penetration, localized heating or overheating near the surface of frozen foods at high intensities (Miles *et al.* 1999; Li & Sun, 2002). Many studies have been performed in order to achieve successful results for thawing frozen foods by varying ultrasound parameters such as frequency and power. It was reported that acceptable ultrasonic thawing was achieved at frequencies around 500 kHz for cod muscle. Also, a model of ultrasonic thawing, predicting the thawing time and heating of a thawed tissue were in close agreement with experimental observations (Miles *et al.* 1999). A frozen block of Pacific cod was exposed to 1500 Hz acoustic energy in an 18°C circulated water bath up to 60 W. The frozen fish block thawed in 71% less time than water-only controls (Kissam *et al.* 1981).

In case of choosing the appropriate frequency and power, ultrasonic thawing is a promising technology.

9.2.9 Summary

There is no one perfect thawer. Each production process will have demands on its own that need to be taken into account. The thawer should meet these goals:

- Give a controllable and predictable process
- Give reproducible thawing runs
- Result in a homogeneous end-temperature
- Minimize the losses
- Give optimal and even quality
- Give short thawing time
- Have little or no environmental impact
- Be energy efficient
- Have high capacity
- Be compact – not space demanding.
- Be flexible, regarding the product size and shape
- Be easy to clean, and suitable for hygienic production
- Be reliable and safe to operate
- Be simple to run and maintain
- Not be labour intensive
- Demand low investments
- Be run continuously/semi-continuously
- Be easy to fit into any production plan (Haugland 2002).

9.3 Recommendation for GMP in seafood thawing

Seafood is a highly perishable food product. It is important to provide the quality and safety of the product from farm to fork. Most hazards related to consumption of fish and fish products can be controlled by applying GMP, good hygienic process (GHP) and hazard analyses in critical control points (HACCP) systems. GMP describes the methods, equipment, facilities, and controls for producing food. Seafood thawing is one of the steps of obtaining safe and good quality fish and fish products. Recommendations to perform GMP in seafood thawing are summarized in Table 9.2 (Archer *et al.* 2008).

Table 9.2 Recommendations for GMP in seafood thawing

	Recommendation
Temperature	• Ensure product temperature is monitored throughout thawing cycle. • Product temperature should be kept close to the temperature of melting ice, i.e. as close to 0°C as possible. • Ensure water and air temperatures are monitored and do not exceed recommended limits. • Allow seafood temperature to equilibrate after thawing, as different parts of the product will be at different temperatures.
Timescale	• Ensure the timescale is appropriate for thawing the seafood. In general, seafood is best thawed quickly (0–6 hours), but not so quickly that product safety and quality is compromised.
Product	• If double freezing (i.e. refreezing thawed seafood) ensure that thawing is done to the highest standards of GMP to reduce risks of changes in texture. • Thawed seafood will spoil as rapidly as chilled never-frozen seafood. Ensure thawed seafood remains at chill temperatures (as close to 0°C as possible). • Do not process under-thawed product, unless it has been specifically tempered for use in another process.
Process and Equipment	• Ensure the process used is monitored and controlled throughout – do not leave the seafood thawing without any supervision as this can lead to under- or overthawing. • Use the most appropriate method and equipment for the product. • Thawing conditions should be clean and hygienic. • Do not use water to thaw cut or processed seafood – only use water to thaw whole or semiprocessed product, e.g. headed and gutted fish. • When purchasing thawing equipment ensure it has proven expertise with seafood products. Ask for demonstrations wherever possible.

References

Alizadeh E, Chapleau N, Delamballerie M. & Lebail A. (2007). Effects of freezing and thawing processes on the quality of atlantic salmon (*Salmo salar*) fillets. *Food Engineering and Physical Properties,* **72**: 279–285.

Archer, M., Edmonds, M. & George, M. (2008). *Seafood Thawing.* Campden and Chorleywood Food Research Association, Research and Development Department, SR598.

Awad, T.S., Moharram, H.A., Shaltout, O.E., Asker, D. & Youssef, M.M. (2012). Applications of ultrasound in analysis, processing and quality control of food: A review. *Food Research International,* **48**: 410–427.

Balaban, M.O., Henderson, T., Teixeira, A. & Otwell, W.S. (1994). Ohmic thawing of shrimp blocks. In: *Developments in Food Engineering,* (Eds Yano, T.; Matsuno, R., & Nakamura, K.), pp. 307–309, Blackie Academic and Professional, London.

Bengtsson, N. (1963). Electronic defrosting of meat and fish at 35 MHz and 2,450 MHz: a laboratory comparison. 5th International Congress on Electro-Heat, Wiesbaden, Germany. Presentation No. 414.

Bermudez-Aguirrre, D., Mobbs, T. & Barbosa-Canovas, G.V. (2011). Ultrasound technologies for food and bioprocessing. In: *Ultrasonic Applications in Food Processing* (Eds Feng H., Barbosa-Cánovas, G., Weiss, J.), pp. 65–106, Springer Publishing, New York.

Bezanson, A., Learson, R. & Teich, W. (1973). Defrosting shrimp with microwaves. *Proceedings of the Gulf and Caribbean Fisheries Institute,* **25**: 44–55.

Bridgman, P. (1912). Water in the liquid and five solid forms under pressure. *Proceedings of the American Academy of Arts and Science,* **47**: 441–558.

Cemeroğlu, B. (2005). *Gıda Mühendisliğinde Temel İşlemler,* pp. 505. Gıda Teknolojisi Derneği Yayınları. No:29, Ankara, Turkey.

Chemat, F., Huma, Z. & Khan, M.K. (2011). Applications of ultrasound in food technology: Processing, preservation and extraction. *Ultrasonics Sonochemistry,* **18** (4): 813–835.

Chevalier, D., Le Bail, A., Chourot, J. & Chantreau, P. (1999). High pressure thawing of fish (Whiting): Influence of the process parameters on drip-losses. *Lebenmittel.- Wissen-schaft und-Technologie,* **32**: 25–31.

Chevalier, D., Sequeira-Munoz, A., Le Bail, A., Simpson, B. & Ghoul, M. (2000). Effect of freezing conditions and storage on ice crystals and drip volume in turbot (*Scophthalmus maximus*), evaluation of pressure shift freezing vs. air-blast freezing. *Innovative Food Science and Emerging Technologies,* **1**: 193–201.

Cybulska, B. & Doe, P.E. (2002). Water and food quality. *Chemical and Functional Properties of Food Components,* pp. 25–49, CRC Press, Boca Raton.

Delbarre-Ladrat, C., Cheret, R., Taylor, R., & Verrez-Bagnis, V. (2006). Trends in postmortem aging in fish: Understanding of proteolysis and disorganization of the myofibrillar structure. *Critical Reviews in Food Science and Nutrition,* **46** (5): 409–421.

Dolatowski, Z.J., Stadnik, J. & Stasiak, D. (2007). Applications of ultrasound in food technology. *Acta Scientiarum Polonorum, Technologies Alimentaria,* **6** (3): 89–99.

Doughikollaee, E.A. (2012). Freezing, thawing and cooking of fish. Scientific, Health and Social Aspects of the Food Industry, Dr. Benjamin Valdez (Ed.)

Available from: <http://www.intechopen.com/books/scientific-health-and-social-aspects-of-the-food-industry/freezing-thawing-and-cooking-of-fish> (accessed 20 January 2015).

Erickson, M.C. & Hung, Y. (1997). *Quality in Frozen Food*, pp. 487, Chapman & Hall, International Thompson Publishing, New York.

Ersoy, B., Aksan, E. & Ozeren, A. (2008). The effect of thawing methods on the quality of eels (*Anguilla anguilla*). *Food Chemistry*, **111**: 377–380.

Gökoğlu, N. (2002). *Su Ürünleri İşleme Teknolojisi*, pp. 157, Su Vakfı Yayınları, Istanbul.

Hale, M. B. & Waters, M. E. (1981). Frozen storage stability of whole and headless freshwater prawns *Machrobranchium rosenbergii*. *Marine Fish Review*, **42**: 18–21.

Haugland, A. (2002). *Industrial thawing of fish to improve quality, yield and capacity.* PhD Thesis, pp. 188. Norwegian University of Science and Technology, Faculty of Engineering Science and Technology, Department of Energy and Process Engineering.

Icier, F., Izzetoglu Turgay, G., Bozkurt, H. & Ober, A. (2010). Effects of ohmic thawing on histological and textural properties of beef cuts. *Journal of Food Engineering*, **99**: 360–365.

Jambrak, A.R. (2012). Application of high power ultrasound and microwave in food processing: Extraction. *Journal of Food Processing Technology*, **3**: 12.

Jason, A.C. (2001). *Thawing Frozen Fish.* Torry Advisory Note no 25, Torry Research Station, Aberdeen, UK.

Keizer, C. (1995). *Fish and Fishery Products. Freezing and chilling of fish*, pp. 287–313, CAB International Press, Wallingford.

Kissam, A., Nelson, R., Ngao, J. & Hunter, P. (1981). Water thawing of fish using low frequency acoustics. *Journal of Food Science*, **47** (1): 71–75.

Knorr, D., Schlueter O. & Heinz, V. (1998). Impact of high hydrostatic pressure on phase transitions of foods. *Food Technology*, **52** (9): 42–5.

Kolbe, E. & Kramer, D. (2007). *Planning for Seafood Freezing*, pp. 112. Alaska Sea Grant College Program, Cooper Publishing, Alaska.

Li, B. & Sun, D. (2002). Novel methods for rapid freezing and thawing of foods- a review. *Journal of Food Engineering*, **54**: 175–182.

Luzuriaga, D.A. & Balaban, M.O. (1996). Electrical conductivity of frozen shrimp and flounder at different temperatures and voltage levels. *Journal of Aquatic Food Product Technology*, **5** (3): 41–63.

Makita, T. (1992). Application of high pressure and thermophysical properties of water to biotechnology. *Fluid Phase Equilibrium*, **76**: 87–95.

Mason, T.J., Paniwnyk, L., Chemat, F. & Vian, M.A. (2011). Ultrasonic food processing. *Alternatives to Conventional Food Processing*, pp. 387–414, RSC Publishing, London.

Miles, C.A., Morley, M.J. & Rendell, M. (1999). High power ultrasonic thawing of frozen foods. *Journal of Food Engineering*, **39**: 151–159.

Mozhaev, V. V., Heremans, K., Frank, J., Masson, P. & Balny, C. (1996). High pressure effects on protein structure and function. *Proteins: Structure, Function and Genetics*, **24** (1): 81–91.

Murakami, T., Kimura, I., Yamagishi, T., Yamashita, M., Sugimoto, M. & Satake, M. (1992). Thawing of frozen fish by hydrostatic pressure. In: Proceedings of the first European seminar on high pressure and biotechnology, La Grande Motte, France (Eds Balny, C., Hayashi, R., Heremans, K. & Masson, P.), pp. 329–331.

Otero, L. & Sanz, P.D. (2003). Modelling heat transfer in high pressure food processing: a review. *Innovative Food Science and Emerging Technologies*, **4**: 121–134.

Pangrle, B.J., Ayappa, K.G., Davis, H.T., Davis, E.A. & Gordon, J. (1991). Microwave thawing of cylinders. *AIChE Journal*, **37**: (12) 1789–1800.

Roberts, J.S. Balaban, M.O., Zimmerman, R. & Luzuriaga D. (1998). Design and testing of a prototype ohmic thawing unit. *Computers and Electronics in Agriculture*, **19** (2): 211–222.

Schubring, R., Meyer, C., Schluter, O., Boguslawski, S. & Knorr, D. (2003). Impact of high pressure assisted thawing on the quality of fillets from various fish species. Innovative *Food Science and Emerging Technologies*, **4**: 257–267.

Sriket, P., Benjakul, S., Visessanguan, W. & Kijroongrojana, K. (2007). Comparative studies on the effect of the freeze–thawing process on the physicochemical properties and microstructures of black tiger shrimp (*Penaeus monodon*) and white shrimp (*Penaeus vannamei*) muscle. *Food Chemistry*, **104**: 113–121.

Srinivasan, S., Xiong, Y. L., Blanchard, S. P. & Tidwell, J. H. (1997). Physicochemical changes in prawns (*Machrobrachium rosenbergii*) subjected to multiple freeze–thaw cycles. *Journal of Food Science*, **62**: 123–127.

Tironi, V., Lamballerie, M. & Le-Bail, A. (2010). Quality changes during the frozen storage of sea bass (*Dicentrarchus labrax*) muscle after pressure shift freezing and pressure assisted thawing. *Innovative Food Science and Emerging Technologies*, **11**: 565–573.

Wagner, W., Saul, A. B. & Pruß, A. (1994). International equations for the pressure along the melting and along the sublimation curve of ordinary water substance. *Journal of Physical and Chemical Reference Data*, **23**: 515 –527.

Yerlikaya, P. & Gökoğlu, N. (2005). Effects of different thawing methods on some characteristics of frozen fish. 35th WEFTA (West European Fish Technologists Association), 19–22 September, Antwerp, Belgium.

Yildiz- Turp, G., Sengun, I.Y., Kendirci, P. & Icier, F. (2013). Effect of ohmic treatment on quality characteristic of meat: A review. *Meat Science*, **93**: 441–448.

Zhu, S., Ramaswamy, H.S. & Simpson, B.K. (2004). Effect of high-pressure versus conventional thawing on color, drip loss and texture of Atlantic salmon frozen by different methods. *Lebensmittel-Wissenschaft & Technologie*, **37**: 291–299.

Index

Note: Page numbers in *italics* refer to Figures; those in **bold** to Tables.

Seafood Chilling, Refrigeration and Freezing: Science and Technology, First Edition.
Nalan Gökoğlu and Pınar Yerlikaya.
© 2015 John Wiley & Sons, Ltd. Published 2015 by John Wiley & Sons, Ltd.